**Interdepartmental Committee for Meteorological Services
and Supporting Research (ICMSSR)**

Committee for Environmental Services, Operations and Research Needs (CESORN)

Joint Action Group for Severe Local Storms Operations (JAG/SLSO)

NATIONAL SEVERE LOCAL STORMS OPERATIONS PLAN

Office of the Federal Coordinator for
Meteorological Services and Supporting Research

8455 Colesville Road, Suite 1500
Silver Spring, Maryland 20910
301-427-2002
www.ofcm.gov

FCM-P11-2010
Washington, DC
November 2010

CHANGE AND REVIEW LOG

Use this page to record changes and notices and reviews.

Change Number	Page Numbers	Date Posted	Initials
1			
2			
3			
4			
5			
6			
7			
8			
9			
10			

Changes are indicated by a vertical line in the margin next to the change.

Review Date	Comments	Initials

FOREWORD

In 1967, the Federal Coordinator for Meteorological Services and Supporting Research received an interagency request to develop the first *National Severe Local Storms Operations Plan (NSLSOP)*. This plan is the 26th version of the NSLSOP and is one of several operations plans produced under the auspices of the Federal Coordinator. This plan supersedes the 2001 version and incorporates significant revisions and changes recommended by the participating agencies through their representatives on the Joint Action Group for Severe Local Storms Operations (JAG/SLSO) of the interagency Committee for Environmental Services, Operations, and Research Needs (C/ESORN).

Because of their intensity, small spatial scale, and tendency for rapid development, severe local storms present a great challenge to both the science of meteorology and to the interagency cooperation required to disseminate warning information rapidly. This plan outlines the responsibilities of the various U.S. Federal agencies that provide meteorological services in observing, forecasting, and warning of severe local storms. It also defines meteorological terms used by the agencies preparing severe local storms forecasts and warnings; identifies operational warning criteria and procedures; and discusses communications, observations, and some public release aspects of warnings for severe local storms.

Additional information describing the warning programs of the participating agencies can be found in the following agencies' documents: Air Force Instruction 15-128, *Aerospace Weather Operations Roles and Responsibilities*; Air Force Manual 15-129, *Aerospace Weather Operations: Processes and Procedures*; Air Force Instruction 10-229, *Responding to Severe Weather Events*; National Weather Service Instruction (NWSI) 10-313, *Special Marine Warnings;* NWSI 10-511, *WFO Severe Weather Products Specification*; NWSI 10-512, *National Severe Weather Products Specification;* NWSI 10-922 *WFO Hydrologic Product Specification;* NWSI 10-801, *Airport Weather Warnings;* NWSI 10-811, *En route Forecasts and Advisories*; Marine Corps Warfighting Publication (MCWP) 3-35.7, *MAGTF (Marine Air-Ground Task Force) Meteorological and Oceanographic Support*; OPNAV INSTRUCTION 3140.24F, *Adverse and Severe Weather Warnings and Conditions of Readiness;* NAVMETOCCOMINST 3140.1L, *United States Navy Meteorological & Oceanographic Support System Manual*; and Department of Homeland Security, *The National Response Plan*.

I am grateful to all the members of the JAG/SLSO (see Appendix F) who dedicated their time, experience, and knowledge to update this plan. The plan is a testimony to the Federal agencies working together to serve and protect the citizens of the United States.

//SIGNED//

Samuel P. Williamson
Federal Coordinator for Meteorological Services
and Supporting Research

Table of Contents

Figures

Tables

CHAPTER 1

RESPONSIBILITIES OF COOPERATING AGENCIES

1.1. General

Cooperation and communication among agencies that provide essential meteorological data, information, and dissemination services are the bases for ensuring that users receive the best possible warnings and forecasts of severe local storms. This coordination is achieved through the activities of the Committee for Environmental Services, Operations, and Research Needs (C/ESORN) and the Joint Action Group for Severe Local Storms Operations (JAG/SLSO) in the Office of the Federal Coordinator for Meteorological Services and Supporting Research (OFCM). The responsible Federal departments and agencies who have promulgated the National Severe Local Storms Operations Plan (NSLSOP) have agreed to arrangements to promote the most effective use of their weather-related assets with respect to severe local storm operations. Between major revisions to this plan, changes will be promulgated by a Change Notice. Once received, the changes should be made to the plan and noted in the Change and Review Log on page iv.

1.2. Scope

The procedures and agreements contained herein apply to all of the 50 United States and the U.S. Territories of Puerto Rico, Virgin Islands, American Samoa, and Guam. The plan defines the roles of the individual agencies participating in the severe local storm warning service when more than one agency is involved in the delivery of service in a specific area. When a single agency is involved in any specific area, that agency's procedures should be contained in internal documents and, to the extent possible, be consistent with the NSLSOP practices and procedures.

1.3. Department of Commerce (DOC) Responsibilities

The DOC, through the National Oceanic and Atmospheric Administration (NOAA), is charged with the overall responsibility to implement a responsive, effective national severe local storms warning service.

1.3.1. National Weather Service (NWS)

The NWS will provide timely dissemination of forecasts, warnings, and all significant information regarding severe local storms to the appropriate agencies, marine and aviation interests, and the general public. Specifically, NWS will provide the following:

- Basic surface, upper air, buoy, and radar observations from its network of observing sites

- Additional observations, when required; these observations will be transmitted to any requesting agency by the appropriate communications technologies

- Basic analyses, forecast charts, and radio facsimile charts through the National Centers for Environmental Prediction (NCEP) Central Operations (NCO), Camp Springs, Maryland

- Severe Local Storm Outlooks and Watch Bulletins through the NCEP Storm Prediction Center (SPC), Norman, Oklahoma

- Dissemination of severe weather and flash flood warnings and statements issued by Weather Forecast Offices (WFO) throughout the United States

- Aviation In-flight Weather Advisories issued through the NCEP Aviation Weather Center (AWC) with aviation responsibilities for periods up to 6 hours for aircraft (civilian and military) and amendments, as appropriate

- A concerted effort to collect and relay Pilot Reports (PIREP) in conjunction with the FAA

- Appropriate public educational materials, concerning the severe local storms/flash flood watch/warning service, and development of community preparedness plans

- Points of contact from SPC and AWC to coordinate with Air Force Weather Agency (AFWA) on backup

1.3.2. National Environmental Satellite, Data, and Information Service (NESDIS)

NESDIS will provide the following services:

- Operate satellite systems capable of providing coverage of selected portions of the United States and adjoining coastal areas

- Receive and respond to requests for coverage of specific areas and times. These requests may come from NCEP, a WFO, or appropriate USAF stations through the NCEP Senior Duty Meteorologist (SDM) in NCO and the NESDIS Satellite Analysis Branch (SAB), according to the NESDIS/NWS Satellite Schedule Coordination and Dissemination Procedures (August 2000).

- Provide appropriate satellite data to authorized research facilities

- Provide multidisciplinary environmental analyses to support disaster mitigation and warning services for U.S. Federal agencies and the international community

1.4. Department of Defense (DOD) Responsibilities

1.4.1. U. S. Air Force (USAF)

Air Force Weather (AFW) is responsible for providing weather support to the USAF, U.S. Army, the Air and Army National Guard, the Air Force and Army Reserve, and other DOD customers throughout the world. AFW will provide the following information and services:

- Basic surface, upper-air, and radar observations from its network of stations making such observations

- Additional observations when required and will make selective, nonsensitive reports available to civilian agencies through existing communications with the Federal Aviation Administration (FAA) or, with prior approval, through direct DOD communications

- A concerted effort to collect and relay all pilot reports (PIREPs)

- Transmission of NWS products for severe weather to Continental U.S. (CONUS) DOD agencies via the USAF communications system

- Through AFWA at Offutt Air Force Base (AFB), Nebraska:

  o Mesoscale model backup to NCEP's NCO during emergency situations, when requested

  o Immediate operations backup to NCEP (SPC and AWC) during emergency situations

- Through Operational Weather Squadrons (OWS) at Barksdale AFB, Louisiana; Scott AFB, Illinois; Davis-Monthan AFB, Arizona; and Hickam AFB, Hawaii; provide weather watch, warning, and advisories for all Air Force and Army installations, including Air and Army National Guard and Air Force and Army Reserve forces in their assigned areas of responsibility

1.4.2   U. S. Army

The active, national guard, and reserve components of the Army rely on the responsible AFW OWS as their primary weather provider, with backup from another OWS. Within the CONUS, the Army will rely on SPC and NWS severe weather products and NOAA Weather Radio (NWR) when its assigned OWS and back-up agencies are unable to provide the support.

1.4.3   U. S. Navy (USN) and U. S. Marine Corps (USMC)

The USN and USMC Meteorological and Oceanographic (METOC) Forecast Centers provide severe local storm warnings in support of the Department of the Navy. Within the conterminous United States and offshore waters, requirements for early warnings of hazardous flying conditions and local destructive phenomena are met by using NWS, AFWA, and Fleet Numerical Meteorology and Oceanography Center (FNMOC) products interpreted by personnel of the Naval Meteorology and Oceanography Command (NAVMETOCCOM) and the Marine Corps METOC Service units. Full use is made of information received from NOAA dissemination sources, as well as other military and civil weather circuits. USN and USMC METOC units will provide the following information and services:

- Basic surface, upper-air, and radar observations, including those taken at sea, from its worldwide network of stations making such observations

- Additional observations when required and make all such reports available to civil agencies through existing communications with the Federal Aviation Administration (FAA) or, with prior approval, through direct DOD communications

- A concerted effort to collect and relay PIREPs

- Limited backup of NCO through FNMOC

1.5 Department of Transportation (DOT) Responsibilities

1.5.1 FAA

The FAA will provide the following information and services:

- Basic surface weather observations from its network of observing sites and radar data, per triagency agreement

- Pre-flight and in-flight pilot weather briefings, within designated airspace, which include Airmen's Meteorological Information (AIRMET), Significant Meteorological Information (SIGMET), Convective SIGMETs, urgent pilot reports (UUA), and Center Weather Advisories (CWA) to pilots on a routine basis

- Dissemination/broadcast of AIRMETs, SIGMETs, Convective SIGMETs, UUAs, CWAs, and other hazardous weather advisories via voice and recorded broadcasts

FAA dissemination broadcasts include the Hazardous In-flight Weather Advisory Service (HIWAS), which is a recorded broadcast available on certain air-to-ground frequencies. HIWAS is updated when conditions warrant. Air traffic controllers advise pilots to monitor HIWAS by broadcasting a HIWAS Alert when the HIWAS has been updated. Other broadcasts include the Transcribed Weather Broadcasts (TWEB) for Alaska, the Telephone Information Broadcast (TIBS), and the VOLMET (meteorological information for aircraft in flight) broadcast for international flights entering U.S. domestic airspace.

1.5.2 Federal Highway Administration (FHWA)

The FHWA will provide the following services:

- Assist in making use of Intelligent Transportation Systems as a means of disseminating severe local storm information to both transportation managers and road users

- Work towards assimilating Road Weather observations into the broader weather observation networks

- Work with vehicle manufacturers and others to explore the collection of surface observations from mobile sources (e.g., cars and trucks)

1.6 Department of Homeland Security (DHS)

1.6.1 Federal Emergency Management Agency (FEMA) Responsibilities

FEMA will provide the following services:

- Develop and maintain communications systems in partnership with NWS to ensure that the emergency management community is provided with access to needed NWS products at no recurring cost

- Operate an interstate hot line telephone system (National Warning System [NAWAS]) that connects FEMA Warning Points and NWS Weather Forecast Offices (WFOs)

- Revise and update a Hazard U.S.-Multi Hazard HAZUS-MH model that can estimate risk, damage, and losses for earthquakes, floods, and hurricane winds, both on an annualized loss basis and on a deterministic basis. (See http://www.fema.gov/plan/prevent/hazus/hz_models.shtm)

1.6.2 U.S. Coast Guard (USCG)

The USCG will provide the following capabilities and services:

- Communications circuits for relay of weather observations to NWS in selected areas

- Coastal broadcast facilities at selected locations for weather forecasts, watches, and warnings

- Personnel, vessel, and communications support to the National Data Buoy Center (NDBC) for development, deployment, and operation of moored environmental data buoy systems

- Surface observations to NWS from its coastal facilities and vessels

1.7 Department of Energy (DOE) Responsibilities

Other than forwarding surface observations to appropriate points of contact during a severe storm, DOE and the National Nuclear Security Administration (NNSA) do not have the capability to provide forecasts or warnings to others. The DOE and NNSA sites/facilities rely on the NWS for forecasts, warnings, and all significant information regarding severe local storms.

DOE and NNSA may provide resources, assets, personnel, and expertise to others following a severe local storm either through the DOE Emergency Assistance Program or through an Energy Emergency response (per DOE Order 151. 1C). These activities may be initiated to support interagency plans; Presidential direction; or State, local, or tribal agreements of mutual aid.

In the event of a severe storm at a DOE or NNSA site, DOE may activate its Headquarters Emergency Operations Center to receive, coordinate, and disseminate emergency information to the White House, other Federal agencies, and other appropriate emergency points of contact.

1.8   Department of the Interior (DOI) Responsibilities

The U.S. Geological Survey will provide water-level and stream-flow data on a near-real-time basis to NWS River Forecast Centers for issuing flash flood warnings at locations throughout the United States.

1.9   Exchange of Data, Products, and Forecasting Techniques between Agencies

There will be a mutual exchange of relevant data and products on the part of all concerned agencies outlined in Chapter 1. SPC and AFWA are the units responsible for preparing centralized severe weather forecasts. These forecast products will be exchanged between them, since AFWA provides limited backup for SPC. The National Severe Storms Laboratory (NSSL), SPC, and NESDIS are actively engaged in developing objective severe weather forecasting and analysis techniques. These organizations will engage, whenever possible, in a joint technique development program and will exchange any objective techniques developed.

CHAPTER 2
DEFINITIONS AND CRITERIA

2.1. General

This chapter defines those common meteorological terms, subject to multiple interpretations, that are used by Federal agencies preparing severe local storms forecasts and warnings. Where possible, the definitions are based on or adapted from the American Meteorological Society (AMS) Glossary of Meteorology.

2.2   Severe Local Storm

A severe local storm is a tornado, waterspout, or thunderstorm with winds of 50 knots (25 m/s or 58 mph) or greater and/or hail one inch (25.4 mm) or greater in diameter at the ground. Significant wind damage (several downed trees) or sightings of large hail or a tornado can help supplement official observations (METAR and SPECI) to determine where severe local storms occur. The only exception over land is for "Airport Weather Warnings" where the hail criterion is ≥ ½ inch in diameter. METAR is an aviation routine weather report issued at hourly or half-hourly intervals. It is a description of the meteorological elements observed at an airport at a specific time. SPECI is an aviation special weather report issued when there is significant deterioration or improvement in airport weather conditions.

2.3   Severe Local Storms Season(s)

Tornadoes and severe thunderstorms may occur anywhere and at any time of the year. The months and location of greatest frequency of severe thunderstorms and tornadoes shift from the southeast United States in the early spring, to the southern and central Plains and lower Midwest during the rest of the spring, and into the northern Plains and upper Midwest during the late spring and early summer. The lowest frequency of occurrence is west of the Rocky Mountains.

2.4   Squall Line

A squall line refers to a line of active thunderstorms, either continuous or with breaks, including contiguous precipitation areas resulting from the existence of the thunderstorms. Squall lines can extend hundreds of miles in length and can be especially disruptive to aviation activities.

2.5   Density/Risk of Severe Thunderstorms

The Storm Prediction Center (SPC) issues daily severe weather outlooks describing forecast coverage of severe thunderstorms across the conterminous United States. Three risk categories (Slight, Moderate, and High) are used to symbolize the coverage and intensity of the expected severe weather. Their definitions are:

- Slight risk—Well-organized severe thunderstorms are expected, but in small numbers and/or low coverage. Depending on the size of the area, approximately 5-29 reports of one inch diameter or larger hail, and/or 5-29 wind events, and/or 3-5 tornadoes would be possible.

- Moderate risk—A potential for a greater concentration of severe thunderstorms than the slight risk, and in most situations, greater magnitude of the severe weather. Within a moderate risk area, at least 30 reports of hail 1 inch diameter or larger, or 6-19 tornadoes, or numerous wind events (at least 30 reports that likely would be associated with a squall line, bow echo, or derecho).

- High risk—A major severe weather outbreak is expected, with a high concentration of severe weather reports and an enhanced likelihood of extreme severe (i.e., violent tornadoes or very damaging convective wind events occurring across a large area). In a high risk area, the potential exists for 20 or more tornadoes, some possibly F2 or stronger, or an extreme wind event potentially causing widespread wind damage and higher-end wind gusts (80+ mph) that may result in structural damage.

2.6 Thunderstorm Intensity Categories

Primary hazards in a thunderstorm are wind, hail, flash flooding, and lightning. Flash flooding and lightning may be mentioned in severe weather watches/warnings, if they will have a significant impact on the general public. The following thunderstorm intensity classes will be used in the forecasting and warning functions of concerned agencies:

- Thunderstorm—Wind gusts less than 50 knots and hail, if any, of less than one inch diameter at the surface.

- Severe Thunderstorm—Thunderstorm-related surface winds (sustained or gusts) of 50 knots or greater and/or surface hail one inch diameter or larger. Wind or hail damage may be used to infer the occurrence/existence of a severe thunderstorm. The word "hail" in a watch bulletin implies hail at the surface as well as aloft, unless a qualifying phrase such as "hail aloft" is used.

NOTE: The USAF uses an additional intensity definition for Moderate Thunderstorm—Wind gusts between 35 and 50 knots, and/or hail, if any, of 1/2 inch diameter up to 3/4 inch diameter at the surface.

2.7 Funnel Cloud

A funnel-shaped cloud of condensation, usually extending from a deep convective cloud, and associated with a violently rotating column of air that is not in contact with the ground.

2.8 Tornado

A violently rotating column of air that is in contact with the ground, either pendant from a cumuliform cloud or underneath a cumuliform cloud, and is often (but not always) visible as a funnel cloud.

2.9 Waterspout

In general, any tornado that occurs over a body of water, consists of an intense columnar vortex (usually containing a funnel cloud), and is connected to a cumuliform cloud.

2.10 Flash Flood

A rapid and extreme flow of high water into a normally dry area or a rapid water-level rise in a stream or creek above a predetermined flood level that begins within 6 hours of the causative event (e.g., intense rainfall over a relatively small area, dam failure, ice jam). However, the actual time threshold may vary in different parts of the country. Ongoing flooding can intensify to flash flooding in cases where intense rainfall results in a rapid surge of rising flood waters.

2.11 Other Warning Criteria

All phenomena, other than those classified as severe local storms, paragraph 2.2, described in the various warnings, bulletins, and advisories should be categorized as "other warning criteria" and are not called severe weather phenomena. Such other warning criteria will be listed separately in the appropriate NOAA/NWS publications.

Table 2.1. Types of NWS Messages—General Categories

Product	Lead time	Certainty of event	Purpose/outcome
Outlook	Hours to a few days	*Possible* occurrence of hazardous weather	Adequate notice to those who need hours or longer lead times
Watch	Less than 1 day for severe weather (e.g., thunderstorms, flash floods, tornadoes) 1-2 days for large-scale events (e.g., winter storms, hurricanes)	Risk of hazardous weather has increased but not necessarily occurred	Notify public to carefully monitor local weather conditions; prepare to protect life/property
Advisory	Hours for large-scale events Minutes for severe weather	A hazardous weather event is occurring, is imminent, or has a very high probability of occurrence	Conditions *may* require immediate action to protect life and property; impacts are generally not as severe

			as with a warning
Warning	Hours for large-scale events Minutes for severe weather	A hazardous weather event is occurring, is imminent, or has a very high probability of occurrence	Conditions *require immediate* action to protect life and property

2.12 Convective SIGMETs

Convective SIGMET bulletins for thunderstorms occurring or expected to occur in lines, for an area of thunderstorms covering at least 3,000 square miles, and/or for severe or embedded thunderstorms, are issued by the Aviation Weather Center (AWC) as scheduled products hourly at 55 minutes after the hour (H+55) and as nonscheduled specials. Special Convective SIGMET bulletins cover the conterminous United States and the adjacent coastal waters. For thunderstorm areas less than 3,000 square miles, a Center Weather Service Unit (CWSU) may issue Center Weather Advisories (CWA) for its area of responsibility.

2.13 Collaborative Convective Forecast Product (CCFP)

The CCFP—a graphical representation of expected convective weather at 2, 4, and 6 hours after issuance time—is produced through a collaborative process involving NWS meteorologists at the AWC, CWSU meteorologists, and airline meteorologists. The CCFP is issued every 2 hours, supports the strategic system-wide planning for the National Airspace System (NAS), and is intended to help reduce traffic-flow disruptions caused by convective weather.

2.14 Special Marine Warnings (SMW)

SMWs are for hazardous over-water events of short duration (up to 2 hours) that are not adequately covered by existing marine warnings and forecasts. These events include convective activity, squalls or wind-shift lines, waterspouts, cold-air funnels, and other localized short-lived phenomena. The nationwide criteria for the SMW issuance are forecast winds of 34 knots and/or hail 3/4 inch or more in diameter and/or a waterspout.

CHAPTER 3
GENERAL OPERATIONS AND PROCEDURES

3.1   General

Every effort has been made to standardize terminology, adopt common definitions, and adjust criteria to a common base; however, each agency has different operational watch and warning criteria that must be met. Although standardization will be used wherever possible in forecasts and warnings, each agency retains the right to specify the forecast, watch, and warning criteria needed to carry out its mission.

3.2   National Weather Service (NWS) Watch/Warning Procedures

3.2.1   General

The NWS has statutory responsibility for providing severe local storms watch and warning service for all 50 States. This responsibility is fulfilled by the National Centers for Environmental Prediction (NCEP) Storm Prediction Center (SPC) and the NWS Weather Forecast Offices (WFOs). NCEP Central Operations (NCO), as the central data processing center for the NWS, issues prognostic charts, discussions, and other forecast data and information that are used by the WFOs, SPC, and NCEP's Aviation Weather Center (AWC) in fulfilling their severe local storm responsibilities.

Geographical Responsibilities

For the conterminous United States (CONUS), the SPC issues Severe Weather Outlooks and Watches. SPC does not issue severe local storm watches for Alaska or Hawaii. The WFOs at Anchorage and Honolulu have the responsibility for maintaining weather watches and issuing warnings as needed for their respective States. Each WFO in the 50 States and the U.S. territories, located in the NWS Pacific and Southern Regions, issues a Thunderstorm Outlook/Hazardous Weather Outlook for a forecast of, and Warnings for an imminent threat of, severe thunderstorms and/or tornadoes.

Watch/Warning Criteria

Any or all of the criteria listed in paragraph 2.2 Severe Local Storms, paragraph 2.5 Density/Risk of Severe Thunderstorms, and paragraph 2.6 Thunderstorm Intensity Categories/Severe Thunderstorms may be mentioned in severe weather watches/warnings to indicate more fully that severe weather is expected. Severe weather watches/warnings that mention tornadoes or waterspouts imply that thunderstorm activity, usually severe, is also expected/occurring.

Outlooks

The SPC issues separate Convective Outlooks for severe weather for Day One, Day Two, and Day Three. Each forecast covers 24 hours (see Appendix A, page A-1 for an example of a

Convective Outlook in both a text and graphical format). The outlook conveys forecasts of expected severe weather coverage as defined in Chapter 2. Additionally, the Day 1 and Day 2 Outlook cover the general thunderstorm outlook when greater than a 10 percent chance of a thunderstorm is forecast. A Day 4-8 Severe Weather Outlook is also issued to indicate areas where a severe potential exists several days in advance. A relatively high confidence for severe weather has to be expected before an area is indicated, given the greater uncertainty. Each WFO issues a Thunderstorm Outlook/Hazardous Weather Outlook to outline the severe convective threats of the day, public watches based on SPC watches, and public warnings for imminent severe thunderstorms and/or tornadoes.

Mesoscale Discussions

The SPC issues Mesoscale Discussions (MDs) to convey to CONUS WFOs, the public, media, and emergency managers the current meteorological reasoning for different types of short-term hazardous weather concerns. MDs are nonscheduled, event-driven products. For severe potential and convective trends, SPC will issue a MD 1 to 2 hours prior to a watch issuance. SPC will also issue a MD for severe weather potential when an area is being monitored for a potential convective watch or when thunderstorm development is potentially severe but will not have enough areal coverage or duration to need a convective watch issuance. MDs are normally issued at least every 2 to 3 hours for each convective watch that is in effect and are focused on mesoscale and storm-scale features within the watch area. For an example of a MD, see Appendix A, page A-2.

The following descriptions are used in the MD "CONCERNING…" line to better identify the reasoning behind the particular SPC discussion of the severe potential:

Watch Unlikely
Watch Possible
Watch Likely
Tornado Watch Likely
Severe Thunderstorm Watch Likely
Watch Needed Soon

Public Watches

SPC Public Watches

The SPC collaborates on watch issuances with the local WFOs and issues watches where severe thunderstorms and/or tornadoes are expected (see Appendix A, page A-3, for an example of a Severe Thunderstorm Watch). A Watch Outline Update (WOU) is also issued that lists all of the counties and, if applicable, the coastal zones that are in the watch issuance (see Appendix A, page A-4, for an example of a WOU). The watch type reflects the anticipated predominant threat. A tornado watch is issued when multiple weak tornadoes or at least one strong tornado is anticipated. A severe thunderstorm watch is issued when hail and/or thunderstorms producing damaging winds or large hail are expected to be the primary threat.

Following the issuance of a severe thunderstorm or tornado watch, SPC issues hourly watch status messages indicating which areas remain under the threat of severe weather (see Appendix A. pg. A-5, for an example of a Watch Status message).

Accompanying the public severe thunderstorm or tornado watch is the aviation version of the same watch, which outlines the watch area for plotting purposes (see Appendix A, page A-6 for an example of an Aviation Watch). This watch also includes the expected maximum hail size, strongest thunderstorm wind gusts, and storm motion vector of severe thunderstorms in the watch area.

WFO Public Watch

Once a watch is issued by the SPC, the WFO issues a Watch County Notification (WCN) listing of the counties within the watch in its area of responsibility (see Appendix A, page A-6 for an example of a Watch County Notification). As the event unfolds, the WCN products are issued by the WFO to clear counties from each watch until the watch has expired. The WCN may also be used to add counties to the watch and to extend the valid time of the watch.

Public Warnings

Tornado Warning (TOR)

Each WFO issues tornado warnings (see Appendix A, page A-8, for an example of a tornado warning) where there is radar or satellite indication and/or reliable spotter reports of a tornado. Valid times are usually 15 to 45 minutes. Warnings are often updated with a severe weather statement while the warning is in effect. The warnings use a 'bullet' format to highlight the most important warning parameters such as type of warning, when a warning is in effect, basis for the warning, and an optional storm-path prediction forecasting the times and locations of the severe weather.

Severe Thunderstorm Warning (SVR)

Each WFO issues severe thunderstorm warnings (see Appendix A, page A-8, for an example of a severe thunderstorm warning) when there is radar or satellite indication and/or reliable spotter reports of wind gusts equal to or in excess of 50 knots (58 mph) and/or hail size of one inch diameter (size of a U.S. quarter-dollar) or larger. Valid times are usually 30 to 60 minutes. Warnings are often updated with a severe weather statement while the warning is in effect. These warnings use a 'bullet' format to highlight the most important warning parameters such as type of warning, when a warning is in effect, basis for the warning, and an optional storm-path prediction forecasting the times and locations of the severe weather.

Flash Flood Warning (FFW)

Each WFO issues a flash flood warning (see Appendix A, page A-9, for an example of a flash flood warning) in any of the following circumstances:

- Flash flooding is imminent or occurring. A dam or levee failure is imminent or occurring.

- A sudden failure of a naturally caused stream obstruction (including debris slide, avalanche, or ice jam) is imminent or occurring.

- Precipitation capable of causing flash flooding is indicated by radar, rain gages, and/or satellite.

- Local monitoring and prediction tools indicate flash flooding is likely.

- A hydrologic model indicates flash flooding for locations on small streams.

- A previously issued flash flood warning needs to be extended in time.

Warnings are often updated with a flash flood statement while the warning is in effect. The warnings use a 'bullet' format to highlight the most important warning parameters such as type of warning, when a warning is in effect, basis for the warning, and locations impacted.

Special Marine Warnings (SMW)

The WFOs issue SMWs for hazardous over-water events of short duration (up to 2 hours) and for events inadequately covered by existing marine warnings and forecasts (see Appendix A, page A-10, for an example of an SMW). These events can include convective activity, squalls or wind shift lines, waterspouts, cold air funnels, and other localized short-lived phenomena.

Convective SIGMETs

The AWC issues Convective SIGMET bulletins both hourly, at 55 minutes past the hour, and as required over CONUS and adjacent coastal waters for areas greater than 3000 square miles (see Appendix A, page A-11, for an example of a Convective SIGMET). Negative bulletins are issued if none of the criteria specified in Chapter 2 is met. Convective SIGMETs alert in-flight interests of the following hazards:

- Tornadoes

- Lines of thunderstorms

- Embedded thunderstorms of any intensity

- Active thunderstorms affecting at least 3000 square miles

- Severe thunderstorms with hail of 3/4 inch diameter or greater or winds 50 knots or greater

National Convective Weather Forecast (NCWF) Product

The NWCF supplements the convective airmen's meteorological information (AIRMET) and SIGMET products. The NCWF product provides forecasts of significant thunderstorm locations 1 hour in the future and is updated every 5 minutes. This product is available via Internet from AWC's operational server: http://aviationweather.gov/products/ncwf/.

Center Weather Service Unit (CWSU) Messages

The CWSU prepares the Center Weather Advisory message for the aviation community which can include severe weather convective SIGMETs for areas less than 3000 square miles.

3.3 DOD Watch/Warning Procedures

3.3.1 USAF

The USAF's Air Force Weather (AFW) provides weather warning support for both the Air Force and Army, including the active and reserve components. All sites receive watches and warnings for thunderstorms, severe thunderstorms, tornadic thunderstorms, and lightning strikes. Warning lead times are standardized to the greatest extent possible; specific mission needs may drive nonstandard lead times and criteria. Standardized lead times are 90 minutes for moderate thunderstorms (those with 1/2 inch diameter hail and/or 35-49 knot winds), 2 hours for severe thunderstorms, 15 minutes for tornadoes, and 30 minutes (prior to a thunderstorm within 5 nm of the installation or airfield) for lightning watches. Lightning warnings are issued when lightning is observed within 5 nm of the installation.

Operational Weather Squadrons (OWS)

AFW's OWSs at Barksdale AFB, Louisiana; Scott AFB, Illinois; Davis-Monthan AFB, Arizona; and Hickam AFB, Hawaii, provide forecast services for designated regions of the United States. These squadrons provide weather warning support for active duty, guard, and reserve Air Force and Army installations within their areas of responsibility. Whenever possible, CONUS OWSs will be included on severe weather conference calls between SPC and the WFOs prior to issuance of a severe weather watch. Figure 3-1 graphically represents the relationship between principal CONUS forecast centers in case of severe weather. Routine collaboration between military and civilian forecast centers will ensure forecast consistency and provide full benefit of meteorological expertise.

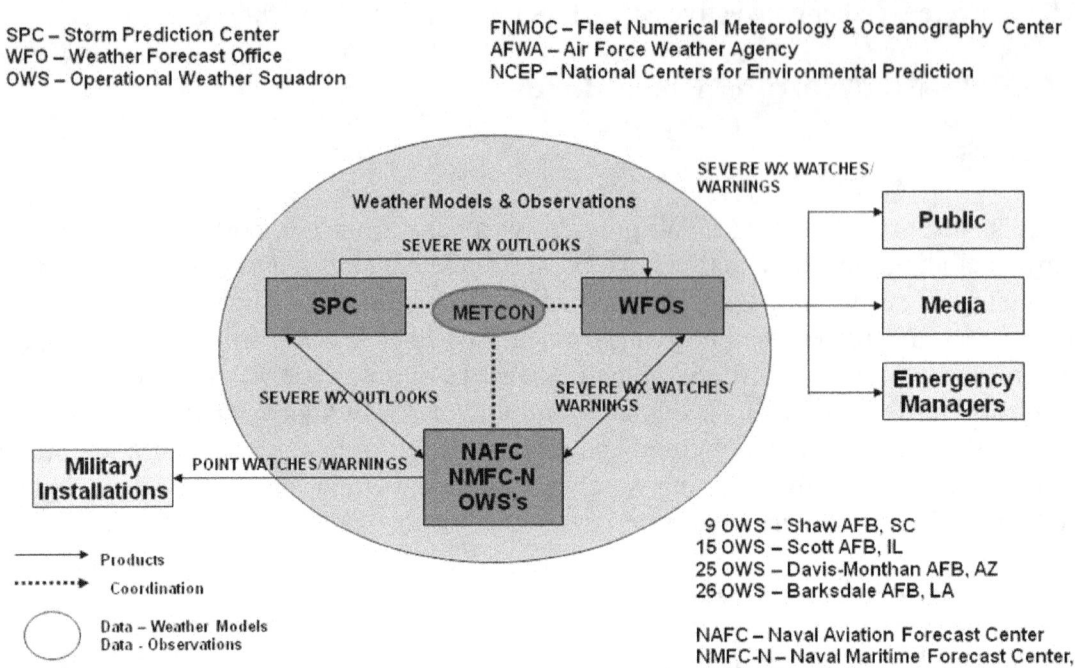

Figure 3-1. Severe Weather Product Generation

Local Unit Warning

At those locations where an AFW forecaster is on duty, the forecaster may issue warnings and advisories, mainly for observed criteria. The criteria and any lead times for warnings are established locally, based on customer needs.

3.3.2 USN and USMC

Navy Meteorology and Oceanography (METOC) Command Centers and Marine Corps weather activities are responsible for the timely dissemination of warnings of hazardous or destructive weather via designated area commands and activities. The Naval Aviation Forecast Center (NAFC) is the centralized aviation weather forecasting hub for CONUS shore-based Naval Aviation activities, providing a combination of 24/7 and after-hours forecasting, severe weather warnings, and flight weather briefing services to 22 Naval Air Stations across the United States. NAFC also exercises operational and administrative control over two Naval Aviation Forecast Detachments and seven Components in Asia and Europe. The Naval Maritime Forecast Center Norfolk (NMFC-N) provides safety of navigation forecasting and route recommendations to a daily average of 70 U.S. Navy, Merchant Marine, and contract carrier vessels operating in the Caribbean Sea, Atlantic Ocean, and Mediterranean Sea. During the Atlantic hurricane season (1 June–30 November), NMFC-N provides advisory services to U.S. Navy shore installations from Brunswick, Maine, to Corpus Christi, Texas, which include National Hurricane Center track forecasts, onset of gale and storm force winds, and storm surge model output for installations affected.

If USN and USMC weather activities are not available, full use should be made of storm warning information disseminated by other agencies (e.g. NWS, USAF, and local foreign meteorological services). In the United States, NWS Bulletins are often heard first over television or radio. Therefore, prior familiarity with their terminology will enhance their value and avoid confusion when Warning Conditions are set by local area commanders. For severe local storms, Conditions II and I are used to avoid or minimize loss and damage due to destructive weather phenomena and are based on all available weather information. Whenever possible, NAFC and NMFC-N will be included on severe weather conference calls between the SPC and the WFOs prior to issuance of a severe weather watch.

3.4 Backup Operations for SPC and AWC

The SPC, AWC, and AFWA have agreed that AFWA will provide limited backup capability for both the SPC and AWC. Appendix C contains the Memorandum of Agreement that covers these backup arrangements, specifying the severe weather forecast and aviation products that AFWA will produce when required to backup either the SPC or AWC. The coordination channel for backup plans and procedures and for exchange of data and products between the SPC, AWC, and AFWA shall be between the Commander, AFWA, and the Director, SPC or AWC, as appropriate. Unresolved differences will be worked out between the Director, NCEP, and the Commander, AFWA

.

CHAPTER 4

COMMUNICATIONS

4.1   Department of Commerce/National Oceanic and Atmospheric Administration (NOAA) Communications Systems

Various distribution methods are used by the National Weather Service (NWS), as appropriate, to make warnings available to NWS field offices, other Federal agencies, National Centers, and the public as rapidly as possible. The NWS Telecommunications Gateway (NWSTG) provides the majority of the connectivity between the producers and users of warnings for these distribution methods. Data collection is also accomplished by several methods within each agency and then shared between agencies. In addition, the NWSTG is the North and South American Regional Meteorological Telecommunications Network (RMTN) for the World Meteorological Organization (WMO) Global Telecommunications System (GTS) which provides global weather data and products to WMO members including the United States. It is also a network of interconnected military, civilian, and foreign computer interfaces used for collecting and distributing environmental data worldwide.

4.1.1   NOAA Weather Wire Service (NWWS)

The NWWS is the primary NWS medium for disseminating warning and forecasts to the media, emergency management agencies, and other users in the public and private sectors. It is a leased satellite communications system operated for the NWS by a private sector contractor. The NWWS will accept messages simultaneously entered from all NWS data entry nodes, primarily NWS Weather Forecast Offices (WFOs) and the National Centers. The system delivers the information to subscribers through a satellite broadcast with output in ASCII format. More information on this system is available via NWS web pages: http://www.nws.noaa.gov/.

4.1.2   NOAA Weather Radio (NWR)

NOAA Weather Radio (NWR) is a nationwide network of over 1000 radio transmitters, broadcasting continuous weather information directly from a nearby NWS office. NWR broadcasts NWS warnings, watches, forecasts, and other hazard information 24 hours a day. In conjunction with the Emergency Alert System, NWR provides an "all-hazards" radio network, making it a single source for comprehensive weather and emergency information. NWR also broadcasts warning and post-event information for all types of hazards: natural (e.g., earthquakes and volcano activity), manmade (such as chemical or environmental incidents), and terrorism-related.

NWS field offices equipped with NWR can transmit continuous weather information on one of following frequencies: 162.400, 162.425, 162.450, 162.475, 162.500, 162.525, and 162.550 MHZ. These radio transmitters provide continuous weather information to an area with a radius of about 40 miles (65 km). Local radio and TV stations can record and rebroadcast the material even when land lines in the area have been disrupted. These transmitters have a tone alert capability used to activate specially designed, commercially available receivers. The NWR network continually broadcasts coastal and marine forecasts. Recorded voice broadcasts are in the process of transitioning to voice synthesis or concatenated voice. The network provides near-continuous coverage of the conterminous United States (CONUS), the Great Lakes, Hawaii, Guam, and the populated Alaska coastline. Typical coverage is 25 nm offshore. A listing of all NWR stations can be found at: http://www.weather.gov/nwr/nwrbro.htm.

NWR 1050 Hz Warning Tone Alarm

An analog 1050 Hz warning alarm precedes many critical watch and warning issuances to activate receivers in a preset muted condition to alert listeners of impending hazards.

NWR Specific Area Message Encoder (SAME)

SAME is a device that puts a special digital code at the beginning and end of selected transmissions of voice messages. The NWS employs SAME with NWR. The SAME code specifies both the type of message (tornado warning, severe thunderstorm watch, etc.) and area (by county) to which the message applies. This gives users, with a decoding device within listening range of the NWR signal, the ability to choose which site-specific hazardous weather messages will automatically interrupt their normal programming. Users of SAME include radio and television stations, schools, cable companies, businesses, and dispatchers. Although SAME will provide much more specificity in both message content and area alerted than the analog 1050 Hz warning alarm, the 1050 Hz warning alarm will continue to be used since it is a long-standing feature of NWR. Many radio manufacturers have designed and developed SAME decoding capability in consumer and industrial grade NWR receivers.

4.1.3   Emergency Managers Weather Information Network (EMWIN)

EMWIN was developed by NWS in partnership with the Federal Emergency Management Agency (FEMA) to ensure access for the emergency management community to a set of NWS products at no recurring cost. The EMWIN data stream contains current weather warnings, watches, images, advisories, and forecasts issued by the NWS.

The information present in the EMWIN data stream originates from WFOs and other sources. This information is collected at the NWS office in Silver Spring, Maryland, then uplinked to the GOES-11 (GOES-West) and GOES-13 (GOES-East) weather satellites from a transmission site in Wallops Island, Virginia.

There are three methods to receive EMWIN data. It can be obtained directly from the GOES satellites using a radio receiver. If the EMWIN signal is being broadcast on a VHF or UHF frequency in a given area, it can be received using a low cost scanner and decoder. It is also available through an Internet connection. Each method requires different hardware and/or software combinations.

EMWIN is continuously broadcast at 9.6 kbps from the GOES-West Satellite (GOES I-M) on a frequency of 1690.725 MHz and at 19.2 kbps on the GOES-East Satellite (GOES N-P) on a frequency of 1692.7 MHz. It contains real-time warnings, watches, advisories, and most of the routine products that are currently on the NWWS system. EMWIN also broadcasts satellite imagery and graphics. More information on EMWIN data content and reception methods is available on the EMWIN website at EMWIN.net.

4.1.4   Low-Rate Information Transmission (LRIT)

The EMWIN data stream is incorporated into the NOAA/NESDIS Low-Rate Information Transmission (LRIT) service provided by GOES-East and GOES-West satellites. This system provides unidirectional broadcast link connectivity between the originating uplink from the NOAA Command and Data Acquisition Stations (CDAS) at Wallops Island, Virginia (WCDAS) and a large number of outlying ground LRIT terminals (LRITT). These LRITTs are typically small receive-only stations.

In addition to EMWIN data, GOES imagery products are generated at the NESDIS Environmental Satellite Processing Center (ESPC) in Suitland, Maryland, and delivered to the CDAS as part of the LRIT data stream for rebroadcast via the GOES satellites. Also included is a copy of the GOES Data Collection System (DCS) data stream and other environmental products including: tropical storm information from the NOAA/National Hurricane Center and Japan Meteorological Agency MTSAT imagery and EUMETSAT Meteosate MSG in graphic format, also produced at the ESPC. The LRIT downlink frequency of 1691 MHz is then converted to an intermediate frequency at the LRITT with a bandwidth that allows a data rate of up to 256 kbps. The GOES system consists of several observing subsystems including the data collection system (DCS). The DCS uses the GOES spacecraft for the relay of data from remotely located insitu sites at or near the Earth's surface.

4.1.5   Really Simple Syndication (RSS)

The NWS leverages multiple technologies to disseminate weather information via the Internet. In addition to the primary NWS web site, www.weather.gov, the NWS has begun to provide RSS feeds at www.weather.gov/rss/. Really Simple Syndication (RSS) is a family of web formats used to publish frequently updated digital content. Users of RSS content use programs called feed 'readers' or 'aggregators' (newer versions of Web browsers offer built in support for RSS feeds): the user 'subscribes' to a feed by entering the link of the RSS feed into their RSS feed reader; the RSS feed reader then checks the subscribed feeds to see if any have new content since the last time it checked, and if so, retrieves the new content and present it to the user.

4.1.6   NOAA Family of Services

NWS provides external users with access to weather information through a collection of data services called the Family of Services (FOS). FOS is accessible via dedicated telecommunications access lines from the Washington, D.C., area. Users may obtain individual services from NWS for a one-time connection charge and an annual user fee. The part of FOS that specifically pertains to forecasts, watches, and warnings is the Public Product Service (PPS). The PPS carries all public warnings and watches, as well as various hydrological, agricultural, and miscellaneous forecasts and products. The Domestic Data Service (DDS) carries basic observations and various aviation, marine, and miscellaneous products.

4.1.7   NOAAPort

The NOAAPort broadcast system provides a one-way broadcast of NOAA environmental data, forecasts, and watch and warning information to NOAA sites and to external users. This service is implemented by a commercial provider using C-band satellite communications. The Advanced Weather Interactive Processing System (AWIPS) Network Control Facility (NCF) routes products and data to the appropriate NOAAPort channels for uplink and broadcast (see Figure 4-1). All products available via FOS and NWWS are also available on NOAAPort, including access to digital NOAA GOES and POES satellite data. Satellite data are passed to NWSTG, NCF, and NOAAPort by NESDIS ESPC in Suitland, Maryland. The NOAAPort User's Page is available online at http://www.nws.noaa.gov/noaaport/html/noaaport.shtml.

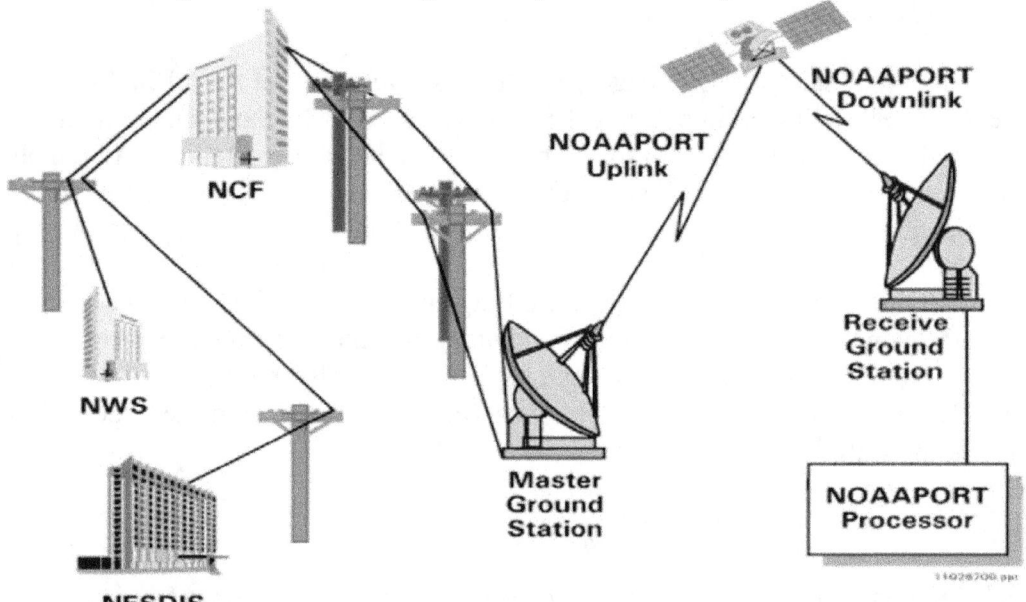

Figure 4-1. NOAAPort Flow of Operational Data and Products

4.1.8   Data Collection and Distribution

Weather data are collected by satellite environmental sensors and surface-based observing systems and processed to create products. Weather data from GOES and POES satellite environmental sensors and Federal agency observations available from NWS WFOs, National Centers, the DOD Automated Weather Network (AWN), and other sources are collected by the NWSTG. As stated previously, satellite data are passed to NWSTG, NCF, and NOAAPort by NESDIS ESPC in Suitland, Maryland. In addition, the NWSTG distributes the data to the nation's operational processing centers and other national and international users through direct links to the NWSTG, the Shared Processing Program (SPP) network, and the Domestic Data Service (DDS). All WFOs have access to the digital GOES satellite data stream through AWIPS workstations. A large amount of satellite data is also available on a number of web-site servers, operated by both governmental agencies and the private sector.

Marine Data Collection Communications

Moored buoy and Coastal Marine Automated Network (C-MAN) data are transmitted by ultrahigh frequency communications via GOES to NESDIS and then relayed to the NWSTG for processing and dissemination. Drifting buoy data are telemetered through the NOAA polar-orbiting satellites to the U.S. Argos Global Processing Center, Largo, Maryland.

Moored buoy observations are formatted into the World Meteorological Organization (WMO) FM 13-IX SHIP code. C-MAN measurements are formatted into C-MAN code, which is similar to the WMO FM 12-IX SYNOP code. The full description of the C-MAN code is contained in the C-MAN Users' Guide, available from National Data Buoy Center (NDBC). Drifting buoy observations are processed and formatted by Service Argos into the WMO FM 18 BUOY code. The messages are then routed to the NWSTG for distribution. Both the SHIP and BUOY codes are defined in the WMO *Manual on Codes*, Volume I.

Radar Products Central Collection/Distribution Services (RPCCDS)

Through the RPCCDS, the AWIPS network collects radar products from NWS, DOD, and FAA wather radar (WSR-88D) sites and delivers them to central radar product collection servers integrated into the NWSTG. All radar products collected are available to users from RPCCDS servers. More information about RPCCDS is available at: http://www.nws.noaa.gov/tg/rpccds.html .

4.2 Department of Homeland Security (DHS)

4.2.1 FEMA Communications System

National Warning System (NAWAS)

NAWAS is the primary system for emergency communications from the Federal government to both State and county warning points. This FEMA-operated, hotline, interstate telephone system connects FEMA warning points with the NOAA/NWS WFOs and National Centers. Figure 4-2 gives the location of FEMA warning points, and Appendix D contains a list of State contacts.

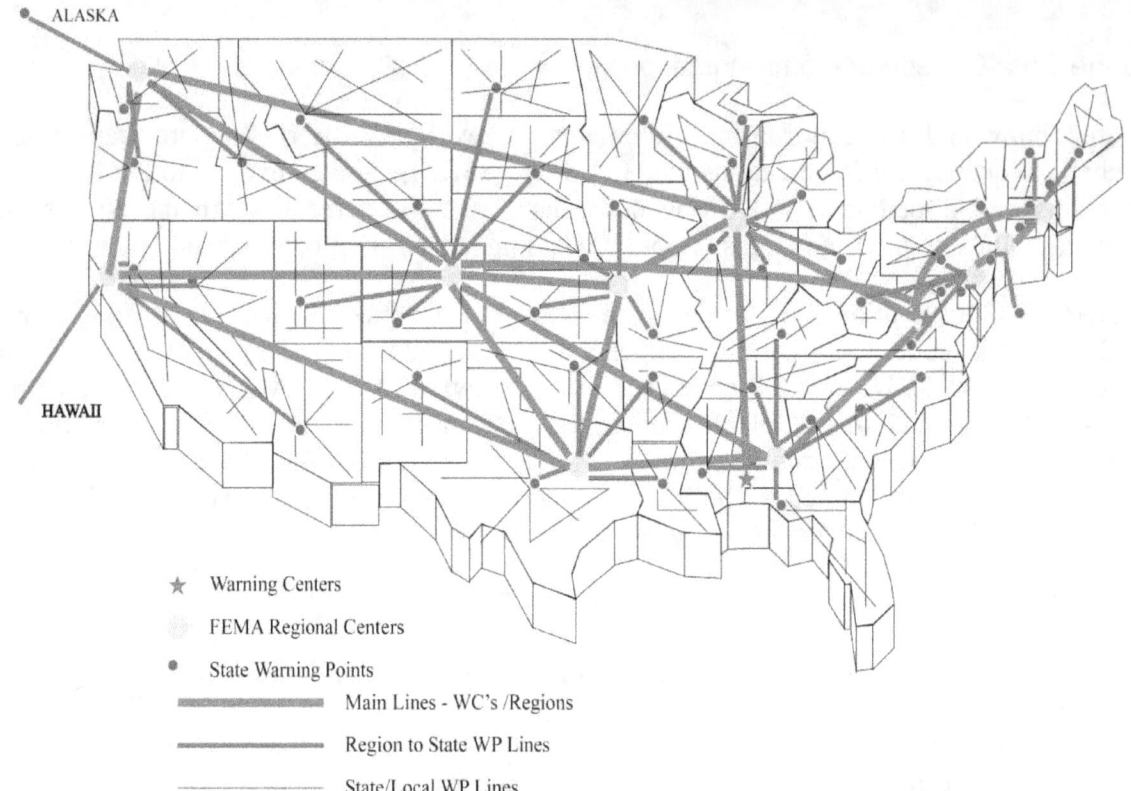

★   Warning Centers

    FEMA Regional Centers

●   State Warning Points

━━━━━   Main Lines - WC's /Regions

━━━━━   Region to State WP Lines

──────   State/Local WP Lines

Figure 4-2. The National Warning System (NAWAS) is FEMA's Operational Hotline Telephone System

4.2.2   U.S. Coast Guard (USCG) Marine Weather Broadcast Systems

The USCG broadcasts forecast, watch, and warning products that include information on severe local storms issued by the NWS National Centers for Environmental Prediction (NCEP) Marine Prediction Center (MPC) and Storm Prediction Center (SPC) and by NWS WFOs. The broadcast of these products supports U.S. participation in the Global Maritime Distress and Safety System, which provides communications support to the International Maritime Organization's global search and rescue plan.

Global Maritime Distress and Safety System (GMDSS)

The goals of GMDSS are to provide more effective and efficient emergency and safety communications, and to disseminate maritime safety information to all ships on the world's oceans, regardless of location or atmospheric conditions. These goals are defined in the International Convention for the Safety of Life at Sea (SOLAS) 1974, as amended in 1988. GMDSS is based upon a combination of satellite and terrestrial radio services and has changed international distress communications from being primarily ship-to-ship based to ship-to-shore (rescue coordination center) based. GMDSS provides for automatic distress alerting and locating, and requires ships to receive broadcasts of maritime safety information that could prevent a distress from happening in the first place. The NWS participates directly in the GMDSS by preparing weather forecasts and warnings for broadcast via two primary GMDSS systems— NAVTEX and Inmarsat-C SafetyNET.

NAVTEX

NAVTEX is an international, automated system for instantly distributing maritime navigational warnings, weather forecasts and warnings, search and rescue notices, and similar information to ships. The International Maritime Organization has designated NAVTEX as the primary means for transmitting urgent coastal marine safety information to ships worldwide. NAVTEX is broadcast from twelve USCG facilities, and coverage is reasonably continuous along the East, West, and Gulf coasts of the United States, as well as in the areas around Kodiak, Alaska; Guam; and Puerto Rico. The typical NAVTEX transmission coverage ranges from 200 to 400 nm.

SafetyNET

Satellite systems operated by Inmarsat, a satellite telecommunications company, offering global mobile services are an important element of the GMDSS. Additional information on Inmarsat can be found at http://www.inmarsat.com/. Inmarsat-C provides ship-to-shore, shore-to-ship, and ship-to-ship store-and-forward data and telex messaging; the capability for sending preformatted messages to a rescue coordination center; and the SafetyNET service. The SafetyNET service is a satellite-based worldwide maritime safety information broadcast service for high-seas weather warnings, navigational warnings, radio navigation warnings, ice reports, warnings generated by the USCG-conducted International Ice Patrol, and other information not provided by NAVTEX.

Coastal Maritime Safety Broadcasts

In addition to NAVTEX and NWR, the USCG and other government agencies broadcast maritime safety information using a variety of radio systems to ensure coverage of different ocean areas for which the United States has responsibility and to ensure ships of every size and nationality can receive this vital safety information.

Very High Frequency (VHF) Marine Radio

The USCG broadcasts near-shore and storm warnings of interest to mariners on VHF channel 22A (157.1 MHZ) following an initial call on the distress, safety, and calling channel 16 (156.8 MHZ). Broadcasts are made from over 200 sites covering the coastal areas of the United States, including the Great Lakes, major inland waterways, Puerto Rico, Alaska, Hawaii, and Guam. All ships over 20 meters in length in U.S. waters are required to monitor VHF channel 16 and must have radios capable of tuning to the VHF simplex channel 22A. Typical broadcast coverage is 25 nm offshore.

Medium Frequency (MF) Radiotelephone (Voice)

The USCG broadcasts offshore forecasts and storm warnings of interest to mariners on 2670 kHz, after first being announced on the distress, safety, and calling frequency 2182 kHz.

Additional Information

Further information concerning these broadcasts can be found at the following Internet sites:

- http://www.navcen.uscg.gov/?pageName=maritimeTelecomms

- http://weather.noaa.gov/fax/marine.shtml

In addition, National Geospatial-Intelligence Agency (NGA) Publication 117 contains detailed information on USCG radio schedules. This publication is available from local National Ocean Service chart agents; it can also be ordered by calling 1-800-638-8972 or 301-436-8301 or by visiting the NOAA Internet site at http://www.nauticalcharts.noaa.gov/staff/chartspubs.html.

4.3 Federal Communications Commission (FCC) Communications System

4.3.1 National Emergency Alert System (EAS)

Formerly known as the Emergency Broadcast System, the National EAS is a nationwide network of broadcast stations and cable systems that provide a readily available and reliable means to communicate emergency information to the American people. State and local authorities have their own EASs, which may also be used to broadcast information on major disasters or emergencies. The FCC designed the National EAS as a tool for officials to quickly send out important emergency information targeted to a specific area. The EAS digital signal uses the SAME coding protocols that the NWS uses on NWR. This allows an NWR signal to be decoded by the EAS equipment at broadcast stations and cable systems, facilitating almost immediate retransmission of NWS weather warning messages to their audiences. The EAS digital system architecture allows broadcast stations, cable systems, participating satellite companies, and other services to send and receive emergency information quickly and automatically even if those facilities are unattended. The National EAS requires monitoring of at least two independent sources for emergency information, to help ensure that emergency information is received and delivered to viewers and listeners. EAS digital messages can be automatically converted into any language used by the broadcast station or cable system or input to external devices used to alert special populations such as the hearing impaired.

4.4 Department of Defense (DOD) Communications Systems

4.4.1 Air Force Communications Systems

Joint Air Force and Army Weather Information Network (JAAWIN) and Joint Environmental Toolkit (JET)

JAAWIN provides access to products via the Internet for any user at a military computer (Internet extension .mil), using digital authentication and encryption technologies. To gain access to the network, nonmilitary users must first request an account and be issued a user name and password. The Internet URL is http://www.afweather.af.mil/.

4.4.2 Navy Communications Systems

Fleet Numerical Meteorology and Oceanography Center (FNMOC)

The U.S. Navy's FNMOC (Internet website https://www.fnmoc.navy.mil/public/) plays a significant role in the National capability for operational weather and ocean prediction through its operation of sophisticated global and regional models whose coverage extends from the top of the atmosphere to the bottom of the ocean. FNMOC is linked with the data collecting and distributing networks of the U.S. Air Force (USAF), NOAA, and WMO. Through these sources, FNMOC collects and assimilates massive volumes of global meteorological and oceanographic (METOC) data for input into its numerical models and distribution to DOD forces worldwide. Utilizing this collection of data, basic and applied computer-generated METOC products are produced for distribution on Navy, Marine Corps, and Joint Command, Control, Communications, Computers, and Intelligence (C^4I) systems.

Many of FNMOC's products are distributed to users over the Internet via the personal computer–based METCAST system and subsequently displayed and manipulated on a user's computer with the Joint METOC Viewer (JMV) software. This includes all standard meteorological and oceanographic fields, synoptic observations, and satellite imagery. For those who require only graphical representation, FNMOC provides a Web-based capability called MyWxMap, which can be accessed through a Web browser for quick display of METOC fields for any user-defined geographical region.

Naval Oceanographic Office (NAVOCEANO)

NAVOCEANO, located at Stennis Space Center, Mississippi, is the primary oceanographic production center for the Navy. It is responsible for collecting, processing, and distributing hydrographic, oceanographic, and other geophysical data and derivative products. Products available from NAVOCEANO include ocean fronts and eddies analyses and surface and three-dimensional ocean thermal fields, which are distributed through the Navy, Marine Corps and Joint C^4I systems.

4.4.3   Data Collection

The Automated Weather Network (AWN) provides the means for data collection within DOD and serves as the DOD link to the WMO GTS through the NWSTG. The AWN currently terminates at AFWA, located at Offutt Air Force Base (AFB). In addition, the High-speed Asynchronous Transfer Mode (ATM) Weather Communications Network (HAWCNET) links Air Force and Navy centers with NOAA's NESDIS and NWS centers to enable sharing of data and products.

Alphanumeric support is provided to end users via the DOD Nonsecure Internet Protocol NETwork (NIPRNET). In addition, the Automatic Digital Network (AUTODIN) via landline, standard DOD C^4I systems, and the Joint Operational Tactical System (JOTS) provides additional means to send METOC data to FNMOC and AFWA and to distribute METOC data and products to users.

4.5   DOT Communications Systems

4.5.1   Federal Aviation Administration (FAA) Systems

Collection of Data and Distribution of Watches, Warnings, and Severe Weather Reports

All FAA air traffic facilities are required to accept and relay pilot reports (PIREPs). FAA satellite, voice, and telecommunications will be used to collect and distribute the following observations and products for severe local storms:

- Routine aviation weather reports (METARs)

- Selected aviation special weather reports (SPECIs)

- PIREPs [Routine pilot reports (UA)/Urgent pilot reports (UUA)]

- Convective Significant Meteorological Information reports (Convective SIGMETs)

- Center Weather Advisories

Weather Message Switching Center Replacement (WMSCR)

WMSCR is the FAA's main weather alphanumeric message switching system. It is designed to store and forward automatically all the various alphanumeric weather messages that contain a proper WMO header. The system consists of two sites, one in Atlanta, Georgia, and the other in Salt Lake City, Utah. These sites normally share the load, but each can support the entire system if the other site is not available.

Automated Flight Service Stations (AFSSs)/Flight Service Stations (FSSs)

The AFSSs/FSSs collect and disseminate PIREPs and broadcast weather information and alerts via air-to-ground radio, telephone recordings, and navigational aids. Requests for information not available at the AFSS/FSS are forwarded to the appropriate NWS office for resolution. These stations also routinely pass information from observers, airport personnel, and pilots to the appropriate NWS office. The FAA and NWS have agreed on the communications methods used to pass this information.

On February 1, 2005, the FAA awarded a contract for the services provided by the 58 AFSSs in CONUS, Puerto Rico, and Hawaii to Lockheed Martin Corporation. Lockheed Martin assumed responsibility for providing AFSS flight services on October 4, 2005. With continued FAA oversight, Lockheed Martin will maintain deliverance of flight services, according to the FAA's strict safety and service requirements. AFSS/FSS in Alaska continues to be operated by the FAA.

4.6   Interagency Shared Processing Program

Polar-orbiting satellite data are processed and exploited by the DOD and NOAA to meet their requirements and are forwarded to each other through the ATM/Shared Processing Program

(SPP) network. The data and products are further distributed to other agencies and the public as appropriate. The ATM/SPP network interconnects the NWSTG with the five U.S. operational processing centers at NCEP, NESDIS, AFWA, FNMOC, and NAVOCEANO. The USAF/USN piece of this SPP connection is HAWCNET (described under DOD systems), which is also used for the exchange of numerical weather prediction model products. The NWS is working to make all of the NOAA polar-orbiting satellite data available over AWIPS. These data are archived on tapes and passed to the National Geophysical Data Center (NGDC) at the University of Colorado for permanent archive.

CHAPTER 5
OBSERVATIONS

5.1 Radar Observing and Reporting Plans

5.1.1 General Description

The Departments of Defense, Commerce, and Transportation operate a national network of Doppler weather surveillance radars designated WSR-88D. Within the Department of Transportation (DOT), the Federal Aviation Administration (FAA) operates three other radar systems: long-range radar, Airport Surveillance Radar (ASR) 9 and 11, and the Terminal Doppler Weather Radar (TDWR).

5.1.2 Observing and Reporting

The WSR-88D Radar Product Generator (RPG) and Terminal Doppler Weather Radar (TDWR) Supplemental Product Generator (SPG) generate graphic products that are distributed to users for detection and evaluation of weather features generally associated with precipitation and storms. The National Weather Service (NWS) centrally collects these graphic products, known as Level III products, from 155 WSR-88D radars and 45 TDWR radars as part of the Radar Product Central Collection Dissemination Service (RPCCDS). Information on how to obtain the products through various methods can be found at http://www.weather.gov/tg/rpccds.html.

Raw radar data, known as Level II data, from most of the NEXRAD WSR-88D radars are now transmitted to the National Centers for Environmental Prediction (NCEP), the National Climatic Data Center (NCDC), and other users (e.g., other government agencies, laboratories, universities, and commercial entities) in real time. The WSR-88D Radar Operations Center is implementing an updated network architecture. The latest information on the WSR-88D Level II Data Collection and Distribution Network, and information on other NEXRAD WSR-88D improvements, can be found at http://www.roc.noaa.gov/WSR88D/

The FAA's long-range radars show reflectivity on displays in the Air Route Traffic Control Center (ARTCC) for en route controllers. An interface was developed that allows these radars to be integrated into AWIPS and is currently used in South Dakota for snow events. However, this interface has potential for severe storms as well. The range is about 200 miles, and the display uses the standard three-color, six-level depiction that is common in the aviation community. The second type of radar equipment includes the ASR 9 and ASR 11, both of which have a weather channel that provides weather information to terminal controllers. This information is used by the Integrated Terminal Weather System (ITWS) to develop mosaics for terminal areas. The mosaics depict microbursts, wind shear, and other current weather parameters, as well as forecasting conditions out to 1 hour. The third radar type, the TDWR, will be installed at 45 locations. It provides information similar to ITWS but not at the level of detail necessary to detect microbursts and forecasts. The FAA intends to eventually make all these data and products available to users, probably through a combination of direct connections and Internet sites. Controllers are encouraged to give pilots the weather information displayed on controllers' radar scopes. The FAA radar product development team (research) is investigating ways to integrate these radars with the WSR-88D system so all the available radar data can be used in detection and forecasting.

5.1.3　National Profiler Network

The NOAA Profiler Network (NPN), consisting of 35 unmanned Doppler radar sites located in 18 central U.S. States and Alaska, provides hourly vertical wind profile data. The data are distributed to the NWS, environmental research groups, and universities. Wind profilers are specifically designed to measure vertical profiles of horizontal wind speed and direction from near the surface to above the tropopause. Profiler locations are concentrated in the central U.S. where most severe weather occurs. See Figure 5-1 for profiler locations. Current information on the NPN can be obtained at http://www.profiler.noaa.gov/npn/.

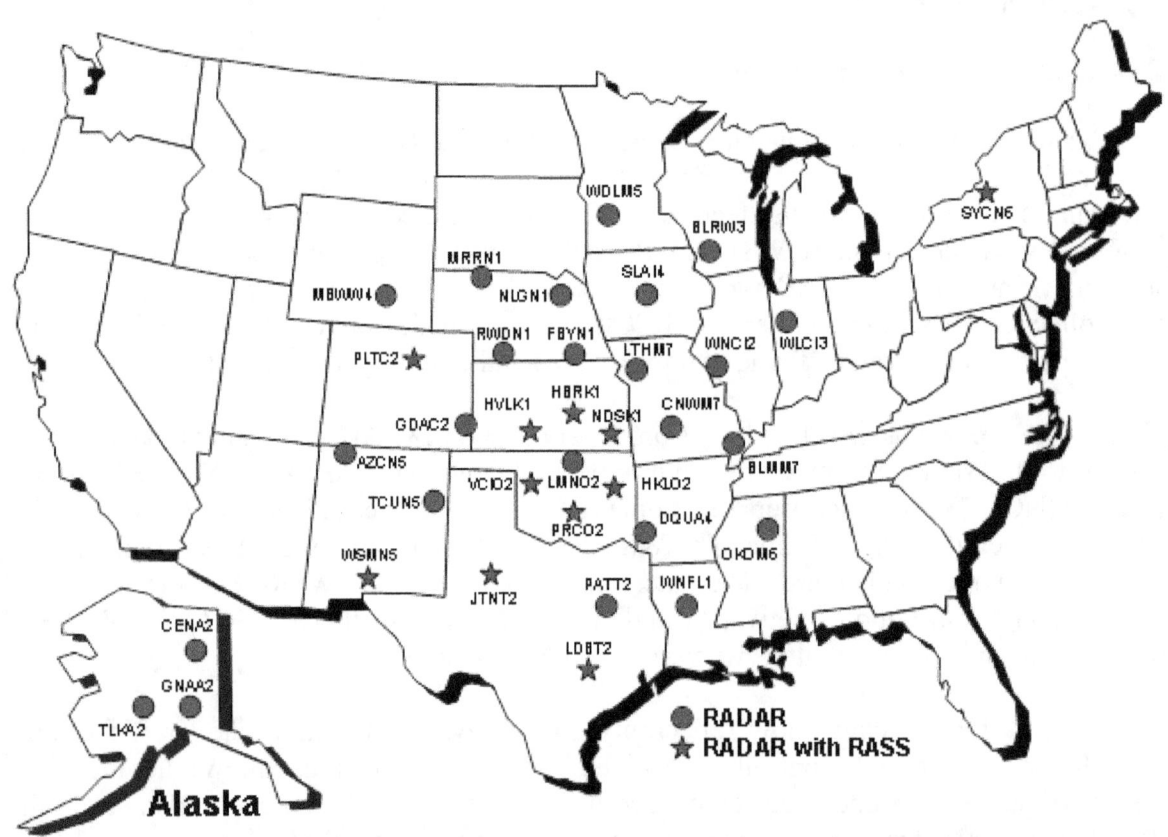

Figure 5-1. Location of National Profiler Network Sites

5.2　Rawinsonde Observing Stations

5.2.1　NWS Network Stations

Rawinsonde observations (RAOB) are made each day at 00:00 UTC and 12:00 UTC at 92 NWS stations—69 in the conterminous United States, 13 in Alaska, 9 in the Pacific, and 1 in Puerto

Rico. See Figure 5-2 for station locations. These stations will take special observations, when requested by the Storm Prediction Center (SPC), in support of severe weather forecasts. Upper-air data from the surface to heights exceeding 30 km are encoded and transmitted to the NWSTG for distribution to Federal agencies and other data users. The NWS Upper-air Observations web page provides further information on the NWS rawinsonde network: http://www.ua.nws.noaa.gov/.

NWS has begun an effort to replace its current network of obsolete rawinsonde observing systems with a modern system that improves the quantity, availability, and accuracy of upper-air data. The new system will utilize Global Positioning System (GPS) radiosondes, which measure winds aloft more accurately than obtained with the current system.

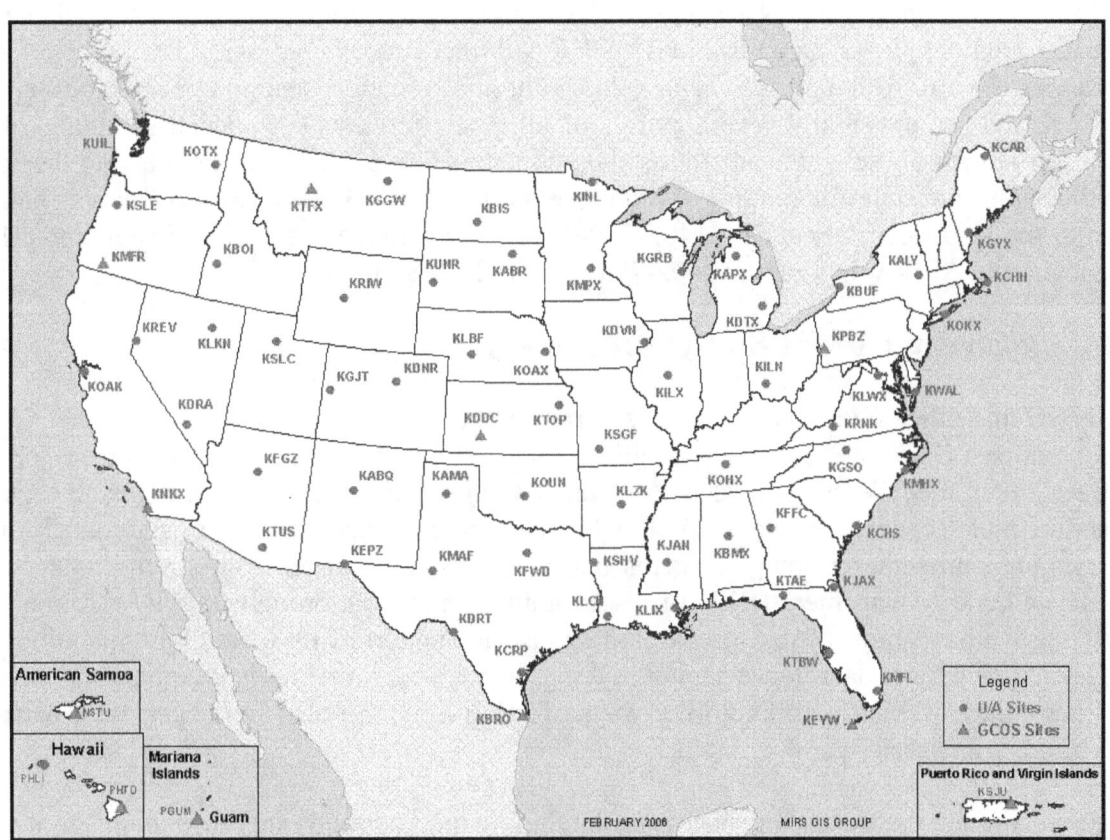

Figure 5-2. Location of RAOB Sites

5.2.2 Other Network Stations

There are approximately five rawinsonde network stations operated by the Department of Defense (DOD), along with additional sites operated by NASA and other Federal agencies, that take and disseminate soundings at 00:00 UTC and/or 12:00 UTC. These sites may not be available to take special soundings in support of severe weather forecasting.

5.2.3   Non-Network Stations

DOD, DOE, and NOAA's Environmental Research Laboratories take unscheduled upper-air observations at mobile locations and Federal facilities in support of weather/climate research programs. Some of these programs encode and disseminate the observations in real time for operational use.

5.2.4   Requests for Special Observations

Any special rawinsonde or pilot balloon (PIBAL) observations needed during the continuous weather monitoring underway at SPC and the Air Force Weather Agency (AFWA) are authorized and will be requested when needed. When special upper-air network soundings are required, the requests should normally be made for 0600Z or 1800Z. The lead forecaster at SPC will initiate the request to the NWS and NASA stations. The Commander, AFWA, will similarly request soundings from DOD stations. SPC will then coordinate with the NCEP Central Operations (NCO). Although WFOs have the authority to request special upper air observations during periods of potentially severe storms of all types, requests for special soundings during periods of potentially severe local storms should be made by SPC. The agency taking the special sounding is responsible for funding. Military requests for NWS or NASA soundings should be made to the lead forecaster at SPC (405-579-0702). NWS requests for USAF soundings should be made to the AFWA duty officer (402-294-2586 or FTS 866-2586).

5.3   Lightning Detection System (LDS)

The NWS and other government agencies currently incorporate lightning data into their day-to-day operations. Since 1996, the Federal government has purchased lightning data from a private contractor with the NWS serving as the contract administrator. The FAA's Automated Lightning Detection and Reporting System (ALDARS) incorporate these data into automated surface observations and some prototype systems that combine information from several sources. The Bureau of Land Management (BLM) uses lightning strike location and polarity information in managing wildland fires. NASA uses lightning data in support of its spaceflight operations. The DOD uses lightning data for decision-making related to refueling and munitions handling activities. These agencies are exploring ways of using lightning data in concert with data from other sources and sensors.

Research continues in the use of space-based lightning mappers and other technology and in the development of capability to depict total lightning strikes as a forecasting and warning tool.

5.4   Surface Weather Observational Network

5.4.1   Land Surface Observations

To provide the basic weather data needed for analyses performed by NCEP, SPC, AFWA, and FNMOC, all available surface data are used. Data are provided through the following sources:

- WFOs/Data Collection Offices (DCOs), Automated Meteorological Observing Stations (AMOS), Supplementary Aviation Weather Reporting Stations (SAWRS), and A-Paid Stations (private contractors paid to take an aviation surface observation).

- DOT/FAA and U.S. Coast Guard (USCG) weather reporting stations, including flight service stations, towers, bases, and contract weather observing stations.

- DOD weather reporting stations.

- Automatic surface observing systems such as the Automated Surface Observing System (ASOS), the Automated Weather Observing System (AWOS), and their replacements.

- Department of the Interior weather and hydrological reporting stations (Remote Automated Weather Station [RAWS] and Snowpack Telemetry Data [SNOTEL]).

- State DOT Environmental Sensor Station observations used in Road Weather Information Systems, available through surface weather observing systems, e.g., Meteorological Assimilation Data Ingest System [MADIS] or *Clarus*.

Augmentation and Backup of Automated Surface Observing Systems

Augmentation is accomplished at staffed locations and is defined as the process of manually observing and adding weather information to an automated surface observing system's observation that the system is not capable of providing. At designated airport locations, NWS and FAA observers are required to augment ASOS observations as defined by the ASOS Service standards listed under each FAA Aviation Service Level (Figure 5-3). At all sites, the minimum required augmentation is to report tornadic activity, hail, virga, volcanic ash, and/or thunderstorms (except where ALDARS is available).

NWS and FAA observers also are required to backup ASOS at designated locations to ensure missing or nonrepresentative data are corrected. Backup is the manual observation and reporting of elements that the system would normally report but are missing or considered not representative. To support daily climatological records, all NWS staffed sites, collocated with an ASOS, are also required to provide other specified data in a daily Supplementary Climatological Data product.

5.4.2   Marine Surface Observations

Marine surface observations are taken by observers at land stations, on ships, and by automated reporting from automated reporting stations. The National Data Buoy Center (NDBC) maintains automated reporting stations in the Gulf of Mexico, in the coastal and offshore areas of the Atlantic and Pacific Oceans, and in the Great Lakes. These data acquisition systems obtain measurements of meteorological and oceanographic parameters for operations and research purposes. Maps showing locations of all buoy and C-MAN stations can be found at: http://www.ndbc.noaa.gov.

"D" Service Level:
No augmentation required - stand alone ASOS

"C" Service Level:
Thunderstorm occurrence
Funnel clouds
Hail
Virga
Volcanic Ash
Tower visibility

"B" Service Level:
all "C" Level required data
Long-line Runway Visual Range (RVR) at designated sites
(RVR 10 minute mean or instantaneous reading)
Freezing drizzle or freezing rain
Ice pellets
Snow depth on ground
Snow increasing rapidly remark (SNOINCR)
Thunderstorm/lightning location remark
Observed significant weather not at station

"A" Service Level:
all "C" and "B" required data
Long-line RVR at designated sites
(RVR 10 minute mean or visibility increments
down to 1/8, 1/16, and 0)
Sector visibility
Variable sky
Cloud types
Cloud layers above 12,000 feet
Widespread dust, sand, and smoke obstructions
Volcanic eruptions

Figure 5-3. Civilian Airports With ASOS. Civilian airports with ASOSs are assigned a specific FAA Aviation Service Level (A, B, C, and D), which has associated ASOS Service standards as specified above. These standards specify what additional data, if any, are required to be observed and added to each ASOS observation.

Personnel on USN, USCG, and NOAA ships, along with civilian Volunteer Observing Ships (VOS) at sea, take and transmit marine observations back to the United States. About 1,600 ships participate in the VOS program, which is managed by NWS, by taking and transmitting the marine observations every 6 hours.

Data Acquisition

Moored buoy and C-MAN stations routinely acquire, store, and transmit data every hour; a few selected stations report every half hour. Data obtained operationally include sea-level pressure, wind speed and direction, peak wind, and air temperature. Sea-surface temperature and wave spectra data are measured by all moored buoys and a limited number of C-MAN stations. Relative humidity is measured at many moored buoy and C-MAN stations where most beneficial to forecast operations.

5.5   Pilot Reports (PIREPs)

5.5.1   Observations

Pilots are encouraged to report weather conditions along the route of flight to confirm forecasted conditions or to indicate conditions differing significantly from the conditions forecasted. Pilots should report any weather conditions they encounter that are hazardous to flight.

5.5.2   Accept/Solicit Reports

All FAA air traffic facilities are required to accept PIREPs. They are also required to solicit PIREPs when current or forecast conditions are below a ceiling of 5,000 feet; visibility is less than 5 miles; and/or when thunderstorms, turbulence, or icing are occurring or forecast. Additionally, Automated Flight Service Stations and Flight Service Stations end all pilot weather briefings with a request for PIREPs.

5.6   Reports by Non-Meteorological Agencies and Individuals

The NWS uses observations of severe local storms, particularly tornadoes, from many non-meteorological agencies and personnel such as: utility companies, State highway patrols, local police departments, road maintenance patrols, citizen spotters (network), cooperative NWS climatological observers, amateur radio groups, local Civil Defense organizations, radio and television station mobile units, city employees, and individual citizens. Local Storm Reports are received by various means and are not uniform at each office. The means include amateur radio or Civil Defense radio facilities with a transceiver often located in the WFOs and operated by local cooperators, police radio, direct telephone lines with unlisted numbers, the National Warning System (NAWAS), State highway patrols, and teletypewriter circuits. Local Storm Reports are disseminated to mass news disseminators, other NWS WFOs, SPC, and safety agencies via NWS circuits (first priority, except for a more expedient means in some local areas). These reports are also verbally disseminated by NAWAS, telephones (hotlines and commercial), and Civil Defense radio facilities. The "fan-out" principle is used wherever practical.

5.7 Severe Storm Surveillance by Meteorological Satellites

5.7.1 Geostationary Operational Environmental Satellite (GOES)

The GOES system consists of two operational spacecraft: GOES-13 (or GOES-East) at 75 degrees west longitude, and GOES-11 (or GOES-West) at 135 degrees west longitude. Upon the successful launch and checkout of GOES-13 in May 2006, GOES-13 replaced GOES-12 in April 2010 and the latter was subsequently moved to 60 degrees west longitude to provide enhanced coverage of South America as part of the Global Earth Observation System of Systems (GEOSS) initiative. The current GOES series (beginning with GOES-8) introduced a 3-axis stabilized geosynchronous satellite to NOAA operations. These satellites ushered in a new era of products and services, providing improved real-time satellite data to the NWS forecast offices and national centers. The GOES-13 satellite provides the same capabilities as GOES-12 and GOES-11, but with a redesigned spacecraft bus for increased navigation and calibration accuracy and increased battery performance, negating the need for eclipse cancellations. GOES-13 represents the first in the next series of GOES satellites, with GOES-14 launched in 2009 and GOES-15 launched in 2010. GOES-14 was successfully checked out and validated in early 2010 and remains on-station at 105 degrees west longitude to provide X-ray sensor data to the Space Weather Prediction Center. GOES-15 completed the NOAA Science Test on September 15, 2010, and is positioned at 90 degrees west longitude as a standby satellite. GOES-11 carries a 5-band imager capable of producing CONUS area images routinely every 15 minutes, at 4 km resolution in the infrared and 1 km in the visible. GOES-13 also carries a 5-band imager but substitutes an 8 km resolution, 13.3 micron band in place of the 12 micron band on GOES-11. The 13.3 micron band is used to measure carbon dioxide emissivity and provides more accurate height measurements of water and ice clouds. The 12 micron band onboard GOES-11 is used in combination with the 10.9 micron band to highlight the emissivity of airborne volcanic ash, which is critical to aviation forecasting. Both satellites employ a 19-band sounder instrument capable of scanning parts of the Northern Hemisphere and CONUS every hour. The sounder provides derived products for storm forecasting, hydrology, fire weather, and input into numerical weather prediction models, as well as providing hourly soundings in clear air.

GOES Scan Operations

The GOES spacecraft routinely scan the United States every 15 minutes, except that every 3 hours a full-disk image is scanned, which takes nearly 30 minutes. Forecasters now view GOES data more frequently and with greater spatial resolution. The GOES-11 and GOES-13 spacecraft were also designed for flexible scanning of the Earth. Any variation of scan or sector coverage at regular time intervals can be scheduled in a 30 minute time frame. Rapid Scan Operations (RSO) and Super Rapid Scan Operations (SRSO) are available on the current generation of GOES satellites. RSO and SRSO operations allow for small sections of the Earth to be scanned more frequently, at up to 1-minute intervals. However by doing so, other portions of the Earth are scanned with less regularity. Definitions of the GOES RSO and SRSO scanning coverage and scanning times can be found at http://www.ssd.noaa.gov. See "GOES Scanning Schedules" on this website.

Requests for Special Satellite Sectors

NWS sites may request, via the NCEP Senior Duty Meteorologist (SDM), RSO and SRSO GOES data on critical severe storm days. The SDM will coordinate this operational request through NESDIS, Satellite Services Division (SSD), Satellite Analysis Branch (SAB). DOD and research requests are taken directly by SAB, which coordinates the request with the NCO SDM.

RSO data are made available to the NWS field offices and NWS National Centers through AWIPS. SRSO data are not made available through AWIPS. During SRSO, AWIPS users will see the standard GOES "routine" schedule.

The details of these procedures are described in the NESDIS/NWS plan, *Satellite Schedule Coordination and Dissemination Procedures*, which is available at the SSD website (http://www.ssd.noaa.gov/PS/SATS/satops/) for users within the government and selected other users (e.g., CIRA and COMET).

Special Products

GOES dissemination schedules for special products are coordinated and provided through NESDIS/SSD and are detailed in the NESDIS/NWS plan, *Satellite Schedule Coordination and Dissemination Procedures*. Visit http://www.ssd.noaa.gov/PS/SATS/satops/ or call 301-763-8444 for more information.

GOES Imagers

GOES-11 and GOES-13 host an imager capable of detecting atmospheric temperature and moisture measurements in five spectral bands at high resolutions, including 3.9 micron and 12.0 micron wavelengths (12.0 micron on GOES-11 only). GOES-11 and GOES-12 also have the feature of transmitting these five spectral bands simultaneously, affording the user community continuous views of atmospheric measurements in various wavelengths, each with its own meteorological and hydrological applications. The five channels and respective resolutions are as follows:

- Channel 1 (visible, 0.55 to 0.75 microns) 1 km resolution

- Channel 2 (infrared, 3.8 to 4.0 microns) 4 km resolution

- Channel 3 (water vapor, 6.5 to 7.0 microns) 4 km resolution on GOES-13 and beyond, 8 km resolution on GOES-11

- Channel 4 (infrared, 10.2 to 11.2 microns) 4 km resolution

- Channel 5 (infrared, 11.5 to 12.5 microns) 4 km resolution (GOES-11 only)

- Channel 6 (infrared, 13.3 microns) 8 km resolution (GOES-13 and beyond only)

GOES Products

The principal GOES-11 and GOES-13 products (see Table 5-2a) are half-hourly pictures with navigation and calibration files included. The most critical products for real-time monitoring of severe storm development are the cloud and moisture imagery indicated as products 1 through 5 in Table 5.2a. During the daylight hours, 1, 2, 4, and 8 km resolution, visible fixed standard sectors are produced for AWIPS/NOAAPort distribution. The infrared sectors (4 km resolution), including both the cloud and water vapor channels (the latter remapped to 4 km) are available 24 hours a day. Satellite raw and remapped imagery, with navigation and calibration, are available to Regional and Mesoscale Meteorological Team Advanced Meteorological Satellite Demonstration and Interpretation System users within the NWS and NESDIS community (see http://www.cira.colostate.edu/cira/RAMM/rmsdsol/main.html for more information), as well as users of the ESPC distribution servers (SATEPSDIST). Products derived from the GOES sounder, including lifted index, land surface temperature, cloud-top pressure, cloud amount, and total precipitable water, are generated hourly at 10 km spatial resolution and provide useful information on trends in large-scale convective activity that could lead to outbreaks of severe weather. Operational and experimental GOES sounder–derived products can be viewed at http://www.ssd.noaa.gov/PS/PCPN/pcpn-na.html#SNDR.

5.7.2   NOAA Polar-Orbiting Satellites

These satellites traverse the United States twice each day at 12hour intervals for each geographical area near the equatorial crossing times listed in Table 5.2a. Data are available via direct readout (HRPT or APT) or central processing. The current primary morning and afternoon polar-orbiting satellites are the Metop-A satellite of the European Organization for the Exploitation of Meteorological Satellites' (EUMETSAT) and the NOAA-19 satellite, respectively, although older satellites still have limited capabilities. The use of the European Metop platforms to fill the primary morning slot is the result of a cooperative effort for data exchange between NOAA and EUMETSAT under the Initial Joint Polar-Orbiting Operational Satellite System (IJPS). In addition to carrying most of the NOAA instruments, the Metop platform hosts instruments with significantly enhanced capabilities for atmospheric sounding (Infrared Atmospheric Sounding Interferometer, or IASI) and marine surface wind vectors (Advanced Scatterometer, or ASCAT). Also, the analog-based APT service has been replaced by the digital-based Low Resolution Picture Transmission (LRPT) service on Metop, which was adopted on NOAA-19. The NOAA-11, NOAA-12, and NOAA-14 platforms were formally decommissioned in June 2004, August 2007, and May 2007, respectively. NOAA-15 and NOAA-16 have been designated as secondary morning and afternoon satellites, and NOAA-18 is designated as a secondary afternoon satellite. NOAA-17 is designated as the morning backup. However, NOAA-17 has an inoperable AMSU instrument, which significantly degrades its capabilities for severe weather monitoring, and NOAA-17's Advanced Very High Resolution Radiometer (AVHRR) has a degraded scan motor that causes periodic noise in the AVHRR images. Daily updates pertaining to the operational status of the various NOAA platforms and the individual instruments can be found at http://www.oso.noaa.gov/poesstatus/.

Polar-Orbiting Environmental Satellite Products

The Polar-Orbiting Environmental Satellite (POES) measurements provide detailed information on the 3-dimensional structure of the atmosphere through the Advanced TIROS Operational Vertical Sounder (ATOVS) package, which consists of the HIRS and AMSU instruments, as well as critical information on bulk cloud and aerosol properties, sea and land surface temperatures, fire and smoke detection, and true color imagery from the high resolution AVHRR instrument. Hydrometeorological parameters such as total precipitable water and cloud liquid water (over ocean), rain rate, ice water path, and snow water equivalent are also available via the multi-channel microwave measurements collected by the AMSU and MHS instruments flown aboard the polar orbiters. In addition, calibrated and navigated radiances from the AMSU and HIRS instruments are provided to NCEP for assimilation into global forecast models. Future plans call for the inclusion of Metop/IASI data. A summary of environmental satellite products generated from NOAA and EUMETSAT polar orbiters is included in Table 5-2a.

Table 5-2a. GOES and NOAA Satellite and Satellite Data Availability for the Severe Local Storms Season

SATELLITE	TYPE OF DATA	LOCAL TIME	PRODUCTS
GOES-13 at 75W(East Ops) GOES-14 at 105W (backup) GOES-15 at 90W (on orbit storage) GOES-11 at 135W (West Ops) GOES-12 at 60W (South America Ops)	Multispectral Imager and Sounder 5 Channels for Imager 19 Channels for Sounder	Every 30 min, in Routine Scan Mode, provides 3 sectors with prescribed coverages: Northern Hemisphere (NH) or Extended NH; Continental U.S. or Pacific U.S.; and Southern Hemisphere (SH). Exception is transmission of full disk every 3 hours. (Available Rapid Scan Operations yield increased transmissions to 7.5 minute intervals to capture rapidly changing, dynamic weather events).	1. 1, 2, 4, and 8 km resolution visible standard sectors. 2. 4 km equivalent resolution IR sectors. 3. Equivalent and full resolution IR enhanced imagery. 4. Full disk Infrared every 3 hours. 5. 4 km water vapor sectors. 6. Clear Sky Brightness Temperatures 7. Quantitative precipitation estimates; high density cloud and water vapor motion wind vectors; and experimental visible and sounder winds. 8. Operational moisture sounder data (precipitable water) in four levels for inclusion in NCEP numerical models. Other sounder products including gradient winds, vertical temperature and moisture profiles, mid-level winds, and derived product imagery (total precipitable water, lifted index, effective cloud amount and surface skin temperature). 9. Tropical storm monitoring and derivation of intensity analysis. 10. Volcanic ash monitoring and dissemination of Volcanic Ash Advisory Statements. 11. Daily northern hemisphere snow cover analysis. 12. Daily fire and smoke analysis over CONUS. 13. Low Cloud / Fog Product

Table 5-2a (cont). GOES and NOAA Satellite and Satellite Data Availability for the Severe Local Storms Season

SATELLITE	TYPE OF DATA	LOCAL TIME	PRODUCTS
Metop-A NOAA-19	AVHRR GAC and LAC (recorded), HRPT & LRPT, AMSU-A, MHS, HIRS, ASCAT, GOME-2, IASI	2130D/0930A (primary morning sat)	1. 1 km resolution HRPT and Local Area Coverage (LAC) data. 2. 4 km resolution APT and Global Area Coverage (GAC) data. 3. Polar Visible and IR mapped imagery.
NOAA-18	AVHRR GAC and LAC (recorded), HRPT & APT, HIRS, AMSU-A, MHS, SBUV-2	0138D/1338A (primary afternoon sat) 0136D/1336A (secondary afternoon sat) 003D/2203A (morning backup)	4. Bulk cloud and aerosol properties 5. Sea-surface temperature analysis. 6. Temperature profiles 7. Moisture profiles.
NOAA-17	Same as NOAA-18 except AMSU-B in place of MHS(AMSU-A inoperable)	0422D/1622A(secondary afternoon sat)	8. Remapped GAC sectors. 9. Sounding-derived products-- total precipitable water, rain rate, cloud liquid water, ice water path, snow water equivalent
NOAA-16	Same as NOAA-17	0510D/1710A (secondary morning sat)	10. Daily northern hemisphere snow cover analysis.
NOAA-15	Same as NOAA-16 except no SBUV-2		11. Twice daily fire and smoke analysis over specific areas within CONUS. 12. Total ozone and stratospheric ozone profiles

AMSU	Advanced Microwave Sounding Unit
GVAR	GOES Variable
LRPT	Low Rate Picture Transmission
(1.1 km) GAC	Global Area Coverage (recorded reduced resolution data for HRPT - High Resolution Picture Transmission (1.1 km) central processing)
APT	Automated Picture Transmission (4 km)
LAC	Local Area Coverage (recorded high-resolution data, limited amount)
AVHRR	Advanced Very High Resolution Radiometer
ATOVS	Advanced TIROS-N Operational Vertical Sounder
SBUV	Solar Backscatter Ultraviolet
MHS	Microwave Humidity Sounder

Under Local Time heading:

D	Descending orbit equator crossing time
A	Ascending orbit equator crossing time

5.7.3 Defense Meteorological Satellite Program (DMSP) Polar-Orbiting System

The DMSP constellation consists of at least two primary operational spacecraft, each placed in sun-synchronous orbits best suited to support military operations (one in an early morning orbit, with equatorial crossing times near the darkness-to-sunlit "terminator" [F-17], and the other with an equatorial crossing time in the mid-morning, near 0830/2030 local time [F-16]). The present constellation also includes several additional secondary operational spacecraft, each with varying capabilities due to degraded sensors, data recorders, command/control systems, etc. In addition to very high-resolution visible and infrared imagery, DMSP provides a variety of remotely sensed terrestrial and space environmental data. A suite of microwave radiometers provides microwave imagery, as well as surface characteristics and upper-air temperature and moisture measurements. Currently, data from the DMSP F-14, F-15, F-16, F-17, and F-18 spacecraft are provided to users However, in 2006, the United States Strategic Command (USSTRATCOM) directed activation of a radar calibration beacon on F-15 that has severely degraded the performance of the 22V GHz channel. This change has in turn impacted the generation of several environmental products derived either directly or indirectly from this channel. The Naval Research Laboratory (NRL) has developed software corrections to mitigate this contamination. Starting with the launch of the first Block 5D3 DMSP spacecraft (F-16) in October, 2003, the capabilities of the SSM/I, SSM/T-1, and SSM/T-2 were combined into a single sensor designated the Special Sensor Microwave Imager/Sounder (SSMIS). DMSP data collection activities are coordinated through AFWA's Second Weather Group request cell (2WXG/DOR). See Table 5-2b.

5.7.4   Shared Processing Program (SPP)

The SPP is a joint Department of Commerce and DOD program whereby NOAA, the Department of the Navy, and the Department of the Air Force cooperate in the acquisition, processing, exchange, and long-term archive of unclassified environmental satellite data and products. Currently, the SPP enables users to access (via NESDIS) information processed by the Navy's Fleet Numerical Meteorology and Oceanography Center (FNMOC) from the SSM/I and SSMIS sensors, including total precipitable water, instantaneous rain rate, soil moisture, snow depth, surface temperature, and ice characteristics. SSM/T-1 data are also made available (via the SPP) at NESDIS for derivation of atmospheric profiles (there are no remaining operational SSM/T-2 sensors)

Table 5-2b. DMSP Satellite and Satellite Data Availability for the Severe Local Storms Season

SATELLITE	TYPE OF DATA	LOCAL TIME (as of 1/22/08)	PRODUCTS
DMSP F-14	OLS Imagery (recorded and direct), SSM/I, SSM/T-1	0505D/1705A	1. 0.3 nm (regional) and 1.5 nm (global) resolution (visible and infrared) imagery available via stored data recovery through AFWA.
DMSP F-15	OLS Imagery (recorded and direct), SSM/I, SSM/T-1, and SSM/T-2 all inop	0714D/1914A	2. Regional coverage at 0.3 nm and 1.5 nm resolution (visible and infrared) imagery available from numerous DOD tactical terminals.
DMSP F-16	OLS Imagery (recorded and direct), SSM/I (22Ghz degraded by radar calibration beacon), SSM/T-1, (SSM/T-2 inop)	0757D/1957A	3. SSM/T-1, SSM/I, SSMIS data transmitted to NESDIS from FNMOC (no remaining SSM/T-2 sensors)
DMSP F-17	OLS Imagery (recorded and direct), SSM/IS imagery	0530D/1730A	
DMSP F-18	OLS Imagery (recorded and direct), SSMIS imagery (derived products currently in cal/val)	0801D/2001A	
	OLS Imagery (recorded and direct), SSMIS imagery (derived products currently in cal/val)		

DMSP	Defense Meteorological Satellite Program
OLS	Operational Linescan Subsystem
SSM/I	Special Sensor Microwave Imager
SSM/IS	Special Sensor Microwave Imager Sounder
SSM/T-1	Special Sensor Microwave Temperature Sounder
SSM/T-2	Special Sensor Microwave Moisture Sounder

Under Local Time heading:

D	Descending orbit equator crossing time
A	Ascending orbit equator crossing time

CHAPTER 6

PUBLICITY

Air Force Weather Agency (AFWA) weather warnings are designed for specialized military users and shall not be released to the public unless provided as backup to the Storm Prediction Center (SPC) or Aviation Weather Center (AWC). News media releases that concern the cooperative efforts in severe local storms activities of the Departments of Defense, Commerce, Transportation, Energy, Homeland Security (Federal Emergency Management Agency), Interior, and other agencies should reflect the joint nature of these efforts by giving due credit to participating agencies. Copies of these releases should be forwarded to:

The Joint Chiefs of Staff (J3/JRC)
Pentagon
Washington, D.C. 20318-3000

HQ USAF/A3O-W
1490 Air Force Pentagon
Washington, D. C. 20330-1490
PH: (702) 696-4021, DSN: 426-4021

HQ AFWA/PA
106 Peace Keeper Suite 2N3
Offutt AFB, NE 68113-4039
PH: (402) 232-8166, DSN: 272-8166

Office of the Deputy Chief of Staff for Intelligence
ATTN: DAMI- POB
2511 Jefferson Davis Highway Suite 9300
Arlington, VA 22202-3910
PH: (703) 601-2499, DSN: 329-2499

Commander, Naval Meteorology and Oceanography Command
1100 Balch Blvd
Stennis Space Center, Mississippi 39529-5005
PH: (228) 688-4203, DSN: 485-4203

Commandant, United States Marine Corps
Headquarters, United States Marine Corps
Code ASL-37
Washington, D. C. 20380-3001

NOAA Public Affairs Office
Herbert C. Hoover Building
14th & Constitution Avenue, N.W.
Washington, D.C. 20230
PH: (202) 482-6090

Public Affairs Officer
NOAA/NESDIS
E/PA, FB-4, Room 3313A
4700 Silver Hill Road
Suitland, MD   20233-0001

Federal Aviation Administration (APA-310)
800 Independence Avenue, S.W.
Washington, DC 20591

Federal Coordinator for Meteorology
Suite 1500
8455 Colesville Road
Silver Spring, MD 20910
PH: (301) 427-2002, DSN: 851-1460

APPENDIX A

TEXT PRODUCT EXAMPLES

Convective Outlook (AC)
```
DAY 1 CONVECTIVE OUTLOOK
NWS STORM PREDICTION CENTER NORMAN OK
0733 AM CDT MON MAY 10 2010

VALID 101300Z - 111200Z

...THERE IS A HIGH RISK OF SVR TSTMS LATE THIS AFTERNOON/EVENING FOR
CENTRAL/NE OK AND SE KS...

...THERE IS A MDT RISK OF SVR TSTMS SURROUNDING THE HIGH RISK IN
KS...OK...SW MO...AND NW AR...

...THERE IS A SLGT RISK OF SVR TSTMS FROM CENTRAL/NW TX TO THE
CENTRAL PLAINS AND MID MS VALLEY THROUGH TONIGHT...

...A FEW STRONG/LONG-TRACK TORNADOES...ALONG WITH VERY LARGE HAIL
AND DAMAGING WINDS...WILL BE POSSIBLE THIS AFTERNOON/EVENING ACROSS
CENTRAL/NE OK AND EXTREME S CENTRAL/SE KS...

...SRN/CENTRAL PLAINS THIS AFTERNOON TO MO OVERNIGHT...
RAPID CHANGES ARE EXPECTED TODAY ACROSS THE SRN/CENTRAL PLAINS AS A
PRONOUNCED SHORTWAVE TROUGH NEAR THE FOUR CORNERS MOVES QUICKLY EWD
TO CENTRAL KS/NRN OK BY THIS EVENING...AND CONTINUES ENEWD TO THE
MID MS VALLEY OVERNIGHT.  A SURFACE WARM FRONT AND ASSOCIATED MOIST
AIR MASS /BOUNDARY LAYER DEWPOINTS OF 65-70 F/ WILL SURGE NWD FROM
NW/N CENTRAL TX TO CENTRAL AND NRN OK BY MID-LATE AFTERNOON.  THIS
MOISTENING WILL OCCUR BENEATH A PLUME OF STEEP MIDLEVEL LAPSE RATES
AND IN RESPONSE TO EWD DEVELOPMENT OF THE SURFACE CYCLONE ACROSS
SW/S CENTRAL KS.  THE MOST FAVORABLE PHASING OF THE UNSTABLE WARM
SECTOR AND STRONG VERTICAL SHEAR ENVIRONMENT WILL OCCUR LATE THIS
AFTERNOON/EVENING ACROSS CENTRAL/N CENTRAL/NE OK...AND THE ADJACENT
BORDER COUNTIES IN KS.  OVERNIGHT...THE BELT OF STRONGER FORCING FOR
ASCENT WILL ACCELERATE ENEWD OVER MO AT A FASTER RATE THAN THE LOW
LEVELS WILL BE ABLE TO MOISTEN/DESTABILIZE...RESULTING IN A DECREASE
IN THE SEVERE STORM/TORNADO RISK LATER TONIGHT.

DISCRETE THUNDERSTORM INITIATION IS EXPECTED BY 18-20Z IN SW KS NEAR
THE SURFACE LOW AND LEFT EXIT REGION OF THE MID-UPPER JET. ADDITIONAL
DEVELOPMENT WILL OCCUR FARTHER SE ALONG THE SRN KS/NRN OK DRYLINE
AROUND 21-22Z...WITH ISOLATED STORMS POSSIBLE SWD INTO CENTRAL OK
AND NW TX CLOSER TO 22-00Z.  THE STRONG DEEP-LAYER SHEAR ENVIRONMENT
NEAR THE MID-UPPER JET CORE SUGGESTS THAT STORMS MAY TAKE UP TO AN
HOUR OR SO TO MATURE.  HOWEVER... EFFECTIVE SRH OF 400-600 M2/S2...
BOUNDARY LAYER DEWPOINTS OF 65-70 F AND MLCAPE OF 2000-3500 J/KG ALL
FAVOR THE POTENTIAL FOR STRONG/LONG-TRACK TORNADOES IN THE MDT-HIGH
RISK AREAS.  THE TORNADO AND LARGE HAIL THREAT SHOULD MAXIMIZE FROM
ABOUT 22-02Z IN OK/SE KS...WITH A GRADUAL EVOLUTION TO A MORE LINEAR
CONVECTIVE MODE WITH DAMAGING WINDS/LARGE HAIL OVERNIGHT ACROSS MO.
```

POTENTIAL CONCERNS WILL BE THE LATE ARRIVAL AND NARROW WIDTH OF THE
WARM SECTOR COMPARED TO THE STRONG VERTICAL SHEAR PROFILES AND FAST
STORM MOTIONS. A COMPARISON OF EXPECTED 45-50 KT STORM MOTIONS WITH
WARM SECTOR WIDTH/PROGRESSION SUGGESTS A 2-3 HOUR WINDOW FOR
SUPERCELLS TO MATURE AND REMAIN WITHIN THE CORRIDOR OF STRONGER
INSTABILITY ACROSS EXTREME SRN/SERN KS AND NRN/NERN OK THIS EVENING.
STILL...IT APPEARS THIS WINDOW OF OPPORTUNITY WILL BE SUFFICIENTLY
LONG TO SUPPORT A SIGNIFICANT SUPERCELL/TORNADO RISK FOR A FEW HOURS
LATE THIS AFTERNOON/EVENING...INCLUDING THE WICHITA...TULSA...AND
OKLAHOMA CITY METROPOLITAN AREAS.

..THOMPSON/HURLBUT.. 05/10/2010

CLICK TO GET WUUS01 PTSDY1 PRODUCT

NOTE: THE NEXT DAY 1 OUTLOOK IS SCHEDULED BY 1630Z

Mesoscale Discussion (MD)

MESOSCALE DISCUSSION 0512
NWS STORM PREDICTION CENTER NORMAN OK
0442 PM CDT MON MAY 10 2010

AREAS AFFECTED...SRN/ERN KS...MUCH OF OK...SWRN MO...NWRN AR

CONCERNING...TORNADO WATCH 147...148...

VALID 102142Z - 102245Z

THE SEVERE WEATHER THREAT FOR TORNADO WATCH 147...148...CONTINUES.

...NEW TORNADO WATCH WILL BE ISSUED SOON FOR PORTIONS OF SERN
KS...SWRN MO...NERN OK...NWRN AR...

RAPID BOUNDARY LAYER RECOVERY IS UNDERWAY ACROSS SCNTRL KS INTO NERN
OK AHEAD OF SFC LOW ALONG THE KS/OK BORDER WEST OF PNC. FOCUSED
PRESSURE FALLS ARE NOTED JUST AHEAD OF THIS FEATURE WITH 2 HOUR
FALLS ON THE ORDER OF 4-7MB. NEEDLESS TO SAY SHEAR AND
THERMODYNAMIC PARAMETERS ARE UNUSUALLY STRONG AND VERY SUPPORTIVE OF
VIOLENT TORNADIC SUPERCELLS. LATEST RADAR TRENDS SUGGEST EARLIER
ISOLATED ACTIVITY IS NOW MORE CONCENTRATED AND FOCUSED ALONG SURGING
DRYLINE WITH AT LEAST HALF A DOZEN STRONG TO RAPIDLY INTENSIFYING
SUPERCELLS...A FEW WITH POTENTIALLY SIGNIFICANT TORNADOES...FROM
KINGMAN COUNTY KS...ARCING TO THE WEST AND SOUTHWEST OF OKC IN SWRN
OKLAHOMA.

LATEST THINKING IS SPEED/MOVEMENT OF CURRENT ACTIVITY WILL REQUIRE
AN ADDITIONAL TORNADO WATCH TO BE ISSUED OVER SERN KS/SWRN MO/NERN
OK/NWRN AR SHORTLY.

..DARROW.. 05/10/2010

ATTN...WFO...SGF...TSA...TOP...ICT...OUN...DDC...

LAT...LON   38909776 38669583 37159468 35329539 34109773 34619858
            36669759 38049856 38909776

Severe Thunderstorm Watch

```
URGENT - IMMEDIATE BROADCAST REQUESTED
SEVERE THUNDERSTORM WATCH NUMBER 678
NWS STORM PREDICTION CENTER NORMAN OK
155 PM CDT TUE SEP 21 2010

THE NWS STORM PREDICTION CENTER HAS ISSUED A
SEVERE THUNDERSTORM WATCH FOR PORTIONS OF

        SOUTHERN IOWA
        FAR WEST-CENTRAL ILLINOIS
        NORTHERN MISSOURI

EFFECTIVE THIS TUESDAY AFTERNOON AND EVENING FROM 155 PM UNTIL
900 PM CDT.

HAIL TO 1.5 INCHES IN DIAMETER...THUNDERSTORM WIND GUSTS TO 70
MPH...AND DANGEROUS LIGHTNING ARE POSSIBLE IN THESE AREAS.

THE SEVERE THUNDERSTORM WATCH AREA IS APPROXIMATELY ALONG AND 65
STATUTE MILES NORTH AND SOUTH OF A LINE FROM 55 MILES NORTH OF
KANSAS CITY MISSOURI TO 20 MILES NORTHEAST OF KIRKSVILLE
MISSOURI.  FOR A COMPLETE DEPICTION OF THE WATCH SEE THE
ASSOCIATED WATCH OUTLINE UPDATE (WOUS64 KWNS WOU8).

REMEMBER...A SEVERE THUNDERSTORM WATCH MEANS CONDITIONS ARE
FAVORABLE FOR SEVERE THUNDERSTORMS IN AND CLOSE TO THE WATCH
AREA. PERSONS IN THESE AREAS SHOULD BE ON THE LOOKOUT FOR
THREATENING WEATHER CONDITIONS AND LISTEN FOR LATER STATEMENTS
AND POSSIBLE WARNINGS. SEVERE THUNDERSTORMS CAN AND OCCASIONALLY
DO PRODUCE TORNADOES.

OTHER WATCH INFORMATION...CONTINUE...WW 676...WW 677...

DISCUSSION...TSTMS ARE EXPECTED TO INTENSIFY THIS AFTERNOON AHEAD OF
WELL-DEFINED VORTICITY MAXIMUM CURRENTLY OVER NERN KS/SERN NEB.
AMBIENT WARM SECTOR AIR MASS IS WARM AND MOIST WITH MLCAPE VALUES OF
1500-2500 J/KG.  WHILE REGION RESIDES ALONG SRN FRINGE OF STRONGER
MID AND HIGH-LEVEL FLOW ATTENDANT TO UPPER GREAT LAKES TROUGH...AREA
PROFILER DATA INDICATE SUFFICIENT DEEP SWLY SHEAR TO SUPPORT
ORGANIZED MULTICELLS AND BOWING STRUCTURES CAPABLE OF DAMAGING WINDS
AND HAIL.

AVIATION...A FEW SEVERE THUNDERSTORMS WITH HAIL SURFACE AND ALOFT
TO 1.5 INCHES. EXTREME TURBULENCE AND SURFACE WIND GUSTS TO 60
KNOTS. A FEW CUMULONIMBI WITH MAXIMUM TOPS TO 500. MEAN STORM
MOTION VECTOR 26030.


...MEAD
```

Watch Outline Update (WOU - Initial for watch number 678 on previous page)
BULLETIN - IMMEDIATE BROADCAST REQUESTED
SEVERE THUNDERSTORM WATCH OUTLINE UPDATE FOR WS 678
NWS STORM PREDICTION CENTER NORMAN OK
155 PM CDT TUE SEP 21 2010

SEVERE THUNDERSTORM WATCH 678 IS IN EFFECT UNTIL 900 PM CDT
FOR THE FOLLOWING LOCATIONS

IAC007-039-051-053-117-135-159-173-179-185-220200-
/O.NEW.KWNS.SV.A.0678.100921T1855Z-100922T0200Z/

IA
.   IOWA COUNTIES INCLUDED ARE

APPANOOSE   CLARKE   DAVIS
DECATUR   LUCAS   MONROE
RINGGOLD   TAYLOR   WAPELLO
WAYNE
$$

ILC001-220200-
/O.NEW.KWNS.SV.A.0678.100921T1855Z-100922T0200Z/

IL
.   ILLINOIS COUNTIES INCLUDED ARE

ADAMS
$$

MOC001-003-021-025-033-041-047-049-061-063-075-079-081-087-089-
095-103-107-111-115-117-121-127-129-137-147-165-171-175-177-195-
197-205-211-227-220200-
/O.NEW.KWNS.SV.A.0678.100921T1855Z-100922T0200Z/

MO
.   MISSOURI COUNTIES INCLUDED ARE

ADAIR   ANDREW   BUCHANAN
CALDWELL   CARROLL   CHARITON
CLAY   CLINTON   DAVIESS
DEKALB   GENTRY   GRUNDY
HARRISON   HOLT   HOWARD
JACKSON   KNOX   LAFAYETTE
LEWIS   LINN   LIVINGSTON
MACON   MARION   MERCER
MONROE   NODAWAY   PLATTE
PUTNAM   RANDOLPH   RAY
SALINE   SCHUYLER   SHELBY
SULLIVAN   WORTH
$$

ATTN...WFO...DMX...EAX...LSX...

Watch Status Message (For Watch 678 on previous page)

```
STATUS REPORT #1 ON WW 678

VALID 212040Z - 212140Z

THE SEVERE WEATHER THREAT CONTINUES ACROSS THE ENTIRE WATCH AREA.

..GRAMS..09/21/10

ATTN...WFO...DMX...LSX...EAX...

&&

STATUS REPORT FOR WS 678

SEVERE WEATHER THREAT CONTINUES FOR THE FOLLOWING AREAS

ILC001-212140-

IL
.    ILLINOIS COUNTIES INCLUDED ARE

ADAMS
$$

IAC007-039-051-053-117-135-159-173-179-185-212140-

IA
.    IOWA COUNTIES INCLUDED ARE

APPANOOSE          CLARKE             DAVIS
DECATUR            LUCAS              MONROE
RINGGOLD           TAYLOR             WAPELLO
WAYNE
$$

MOC001-003-021-025-033-041-047-049-061-063-075-079-081-087-089-
095-103-107-111-115-117-121-127-129-137-147-165-171-175-177-195-
197-205-211-227-212140-

MO
.    MISSOURI COUNTIES INCLUDED ARE

ADAIR              ANDREW             BUCHANAN
CALDWELL           CARROLL            CHARITON
CLAY               CLINTON            DAVIESS
DEKALB             GENTRY             GRUNDY
HARRISON           HOLT               HOWARD
JACKSON            KNOX               LAFAYETTE
LEWIS              LINN               LIVINGSTON
MACON              MARION             MERCER
MONROE             NODAWAY            PLATTE
PUTNAM             RANDOLPH           RAY
SALINE             SCHUYLER           SHELBY
SULLIVAN           WORTH
$$
```

THE WATCH STATUS MESSAGE IS FOR GUIDANCE PURPOSES ONLY. PLEASE
REFER TO WATCH COUNTY NOTIFICATION STATEMENTS FOR OFFICIAL
INFORMATION ON COUNTIES...INDEPENDENT CITIES AND MARINE ZONES
CLEARED FROM SEVERE THUNDERSTORM AND TORNADO WATCHES.

Aviation Watch Notification Message (For Watch #678 above)
WW 678 SEVERE TSTM IA IL MO 211855Z - 220200Z
AXIS..65 STATUTE MILES NORTH AND SOUTH OF LINE..
55N MKC/KANSAS CITY MO/ - 20NE IRK/KIRKSVILLE MO/
..AVIATION COORDS.. 55NM N/S /38N MKC - 17NE IRK/
HAIL SURFACE AND ALOFT..1.5 INCHES. WIND GUSTS..60 KNOTS.
MAX TOPS TO 500. MEAN STORM MOTION VECTOR 26030.

LAT...LON 40849459 41259227 39369227 38979459

THIS IS AN APPROXIMATION TO THE WATCH AREA. FOR A
COMPLETE DEPICTION OF THE WATCH SEE WOUS64 KWNS
FOR WOU8.

Watch County Notification (from WFO in Watch #678 above)
WATCH COUNTY NOTIFICATION FOR WATCH 678
NATIONAL WEATHER SERVICE KANSAS CITY/PLEASANT HILL MO
156 PM CDT TUE SEP 21 2010

MOC001-003-021-025-033-041-047-049-061-063-075-079-081-087-089-
095-107-115-117-121-129-147-165-171-175-177-195-197-211-227-
220200-
/O.NEW.KEAX.SV.A.0678.100921T1856Z-100922T0200Z/

THE NATIONAL WEATHER SERVICE HAS ISSUED SEVERE THUNDERSTORM WATCH
678 IN EFFECT UNTIL 9 PM CDT THIS EVENING FOR THE FOLLOWING AREAS

IN MISSOURI THIS WATCH INCLUDES 30 COUNTIES

IN CENTRAL MISSOURI

HOWARD   SALINE

IN NORTH CENTRAL MISSOURI

CALDWELL   CARROLL   CHARITON
DAVIESS   GRUNDY   HARRISON
LINN MO   LIVINGSTON   MACON
MERCER   PUTNAM   RANDOLPH
SULLIVAN

IN NORTHEAST MISSOURI

ADAIR   SCHUYLER

IN NORTHWEST MISSOURI

ANDREW   BUCHANAN   CLINTON
DEKALB   GENTRY   HOLT
NODAWAY   WORTH

IN WEST CENTRAL MISSOURI

CLAY	JACKSON	LAFAYETTE
PLATTE	RAY	

THIS INCLUDES THE CITIES OF...ALBANY...BETHANY...BROOKFIELD...
CAMERON...CARROLLTON...CHILLICOTHE...CONCORDIA...
EXCELSIOR SPRINGS...FAYETTE...GALLATIN...GRANT CITY...
INDEPENDENCE...KANSAS CITY...KEYTESVILLE...KINGSTON...
KIRKSVILLE...LA PLATA...LANCASTER...LEXINGTON...LIBERTY...MACON...
MARSHALL...MARYVILLE...MILAN...MOBERLY...MOUND CITY...
NEW FRANKLIN...OREGON...PARKVILLE...PLATTE CITY...PLATTSBURG...
PRINCETON...RICHMOND...SAVANNAH...ST. JOSEPH...
TRENTON AND UNIONVILLE.

Tornado Watch (Particularly Dangerous Situation)
URGENT - IMMEDIATE BROADCAST REQUESTED
TORNADO WATCH NUMBER 147
NWS STORM PREDICTION CENTER NORMAN OK
125 PM CDT MON MAY 10 2010

THE NWS STORM PREDICTION CENTER HAS ISSUED A
TORNADO WATCH FOR PORTIONS OF

      SOUTH CENTRAL KANSAS
      CENTRAL AND WESTERN OKLAHOMA

EFFECTIVE THIS MONDAY AFTERNOON AND EVENING FROM 125 PM UNTIL
1000 PM CDT.

...THIS IS A PARTICULARLY DANGEROUS SITUATION...

DESTRUCTIVE TORNADOES...LARGE HAIL TO 4 INCHES IN DIAMETER...
THUNDERSTORM WIND GUSTS TO 80 MPH...AND DANGEROUS LIGHTNING ARE
POSSIBLE IN THESE AREAS.

THE TORNADO WATCH AREA IS APPROXIMATELY ALONG AND 95 STATUTE
MILES EAST AND WEST OF A LINE FROM 40 MILES SOUTH SOUTHEAST OF
FORT SILL OKLAHOMA TO 25 MILES NORTH NORTHEAST OF WICHITA KANSAS.
FOR A COMPLETE DEPICTION OF THE WATCH SEE THE ASSOCIATED WATCH
OUTLINE UPDATE (WOUS64 KWNS WOU7).

REMEMBER...A TORNADO WATCH MEANS CONDITIONS ARE FAVORABLE FOR
TORNADOES AND SEVERE THUNDERSTORMS IN AND CLOSE TO THE WATCH
AREA. PERSONS IN THESE AREAS SHOULD BE ON THE LOOKOUT FOR
THREATENING WEATHER CONDITIONS AND LISTEN FOR LATER STATEMENTS
AND POSSIBLE WARNINGS.

OTHER WATCH INFORMATION...CONTINUE...WW 146...

DISCUSSION...A DANGEROUS ENVIRONMENT IS DEVELOPING ACROSS PARTS OF
SOUTHERN KS AND WESTERN/CENTRAL OK THIS AFTERNOON AS AIRMASS RAPIDLY
DESTABILIZES AND UPPER TROUGH APPROACHES. THUNDERSTORMS ARE
EXPECTED TO INITIALLY DEVELOP ALONG THE SURFACE DRYLINE OVER WESTERN
OK...THEN PROGRESSING EASTWARD THROUGH THE LATE AFTERNOON AND
EVENING. TORNADIC SUPERCELLS ARE A DISTINCT POSSIBILITY WITH THE

THREAT OF STRONG AND LONG-TRACK TORNADOES. VERY LARGE HAIL IS ALSO
LIKELY IN THE STRONGER CELLS.

AVIATION...TORNADOES AND A FEW SEVERE THUNDERSTORMS WITH HAIL
SURFACE AND ALOFT TO 4 INCHES. EXTREME TURBULENCE AND SURFACE
WIND GUSTS TO 70 KNOTS. A FEW CUMULONIMBI WITH MAXIMUM TOPS TO
500. MEAN STORM MOTION VECTOR 25045.

...HART

Tornado Warning (TOR)

BULLETIN - EAS ACTIVATION REQUESTED
TORNADO WARNING
NATIONAL WEATHER SERVICE AMARILLO TX
712 PM CDT MON MAY 31 2010

THE NATIONAL WEATHER SERVICE IN AMARILLO HAS ISSUED A

* TORNADO WARNING FOR...
  NORTH CENTRAL CIMARRON COUNTY IN THE PANHANDLE OF OKLAHOMA.

* UNTIL 745 PM CDT

* AT 711 PM CDT...NATIONAL WEATHER SERVICE METEOROLOGISTS WERE
  TRACKING A LARGE AND EXTREMELY DANGEROUS TORNADO 24 MILES NORTH
  OF BOISE CITY...MOVING SOUTHEAST AT 20 MPH.

* THE TORNADO WILL REMAIN OVER MAINLY RURAL AREAS OF NORTHERN
  CIMARRON COUNTY.

PRECAUTIONARY/PREPAREDNESS ACTIONS...

A TORNADO HAS BEEN CONFIRMED! TAKE COVER IN A STURDY BUILDING NOW.
MOBILE HOMES AND VEHICLES ARE NOT SAFE.

&&

LAT...LON 3699 10223 3694 10215 3677 10259 3700 10267
TIME...MOT...LOC 0011Z 336DEG 17KT 3705 10253

Severe Thunderstorm Warning (SVR)

BULLETIN - EAS ACTIVATION REQUESTED
SEVERE THUNDERSTORM WARNING
NATIONAL WEATHER SERVICE QUAD CITIES IA IL
312 PM CDT TUE SEP 21 2010

THE NATIONAL WEATHER SERVICE IN THE QUAD CITIES HAS ISSUED A

* SEVERE THUNDERSTORM WARNING FOR...
  CENTRAL JO DAVIESS COUNTY IN NORTHWEST ILLINOIS...
  STEPHENSON COUNTY IN NORTHWEST ILLINOIS...

* UNTIL 415 PM CDT.

* AT 310 PM CDT...NATIONAL WEATHER SERVICE DOPPLER RADAR INDICATED A
  SEVERE THUNDERSTORM CAPABLE OF PRODUCING DAMAGING WINDS IN EXCESS

OF 60 MPH....AND PING PONG BALL SIZE HAIL. THIS STORM WAS LOCATED
NEAR WOODBINE...OR 13 MILES SOUTHEAST OF GALENA...AND MOVING EAST
NORTHEAST AT 50 MPH.

* THE SEVERE THUNDERSTORM WILL BE NEAR...
  STOCKTON AND PLUM RIVER AROUND 320 PM CDT...
  LENA AROUND 335 PM CDT...
  CEDARVILLE AROUND 345 PM CDT...
  DAKOTA AROUND 350 PM CDT...
  ROCK CITY AROUND 355 PM CDT...
  DAVIS AROUND 400 PM CDT...

PRECAUTIONARY/PREPAREDNESS ACTIONS...

THIS IS A VERY DANGEROUS STORM CAPABLE OF BLOWING DOWN TREES AND
POWER POLES. MOVE INDOORS TO A SUBSTANTIAL SHELTER SUCH AS A BASEMENT
OR INTERIOR ROOM OR HALLWAY.

A SEVERE THUNDERSTORM WATCH REMAINS IN EFFECT UNTIL 900 PM CDT
TUESDAY EVENING FOR EASTERN IOWA AND NORTHWESTERN ILLINOIS AND
NORTHEAST MISSOURI.

&&

LAT...LON 4249 8940 4225 8940 4224 9026 4238 9028
      4238 8992 4251 8992 4250 8942
TIME...MOT...LOC 2012Z 260DEG 43KT 4235 9016

Flash Flood Warning (FFW)

BULLETIN - EAS ACTIVATION REQUESTED
FLASH FLOOD WARNING
NATIONAL WEATHER SERVICE DES MOINES IA
1110 AM CDT TUE SEP 21 2010

THE NATIONAL WEATHER SERVICE IN DES MOINES HAS ISSUED A

* FLASH FLOOD WARNING FOR...
  MARSHALL COUNTY IN CENTRAL IOWA...
  TAMA COUNTY IN CENTRAL IOWA...

* UNTIL 500 PM CDT

* AT 1109 AM CDT...NATIONAL WEATHER SERVICE DOPPLER RADAR INDICATED
  SLOW MOVING THUNDERSTORMS WITH VERY HEAVY RAINFALL ACROSS THE
  WARNED AREA. RADAR ESTIMATES THAT THESE STORMS HAVE ALREADY
  PRODUCED AS MUCH AS 2 TO 3 INCHES OF RAINFALL OVER THE PAST HOUR.
  AS MUCH AS 1 TO 2 INCHES OF ADDITIONAL RAINFALL IS EXPECTED OVER
  THE NEXT COUPLE OF HOURS.

* RUNOFF FROM THIS EXCESSIVE RAINFALL WILL CAUSE FLASH FLOODING TO
  OCCUR. SOME LOCATIONS THAT WILL EXPERIENCE FLOODING INCLUDE...
  DYSART...MARSHALLTOWN...STATE CENTER AND TOLEDO.

PRECAUTIONARY/PREPAREDNESS ACTIONS...

EXCESSIVE RUNOFF FROM HEAVY RAINFALL WILL CAUSE FLOODING OF SMALL

CREEKS AND STREAMS...HIGHWAYS AND UNDERPASSES. ADDITIONALLY...
COUNTRY ROADS AND FARMLANDS ALONG THE BANKS OF CREEKS...STREAMS AND
OTHER LOW LYING AREAS ARE SUBJECT TO FLOODING.

DO NOT DRIVE YOUR VEHICLE INTO AREAS WHERE THE WATER COVERS THE
ROADWAY. THE WATER DEPTH MAY BE TOO GREAT TO ALLOW YOUR CAR TO CROSS
SAFELY. TURN AROUND...DONT DROWN.

&&

LAT...LON 4187 9323 4217 9323 4221 9303 4221 9278
        4226 9277 4230 9253 4229 9230 4202 9230

Special Marine Warning (SMW)

BULLETIN - IMMEDIATE BROADCAST REQUESTED
SPECIAL MARINE WARNING
NATIONAL WEATHER SERVICE BROWNSVILLE TX
936 AM CDT TUE SEP 21 2010

THE NATIONAL WEATHER SERVICE IN BROWNSVILLE HAS ISSUED A

* SPECIAL MARINE WARNING FOR...
  COASTAL WATERS FROM PORT MANSFIELD TX TO THE RIO GRANDE RIVER OUT
  20 NM.

* UNTIL 1100 AM CDT

* AT 935 AM CDT...NATIONAL WEATHER SERVICE DOPPLER RADAR INDICATED A
  STRONG THUNDERSTORM...WITH WIND GUSTS OF 34 KNOTS OR GREATER...6
  MILES EAST OF SOUTH PADRE ISLAND...MOVING WEST AT 10 KNOTS.

* THE THUNDERSTORM WILL BE NEAR...
  SOUTH PADRE ISLAND BY 1015 AM CDT.

PRECAUTIONARY/PREPAREDNESS ACTIONS...

STRONG THUNDERSTORMS CAN PRODUCE WATERSPOUTS WITH LITTLE OR NO
ADVANCE WARNING. SEEK SAFETY IMMEDIATELY!

MARINERS CAN ALSO EXPECT LOCALLY HIGH WAVES...DANGEROUS LIGHTNING...
AND TORRENTIAL RAIN. SEEK SAFETY IMMEDIATELY.

PLEASE REPORT WATERSPOUTS OR FUNNEL CLOUDS...WINDS OF 34 KNOTS OR
HIGHER...HAIL THE SIZE OF PENNIES OR LARGER...AND ANY VESSEL DAMAGE
TO YOUR NATIONAL WEATHER SERVICE IN BROWNSVILLE BY CALLING
956-504-1432.

&&

LAT...LON 2596 9717 2598 9717 2604 9718 2607 9717
        2620 9719 2616 9688 2595 9688
TIME...MOT...LOC 1435Z 098DEG 7KT 2610 9708

Convective SIGMETs

```
CONVECTIVE SIGMET 38C
VALID UNTIL 2255Z
WI IL MO IA KS NE
FROM 50SW BAE-20W BDF-50S DSM-50W MCI
LINE SEV TS 25 NM WIDE MOV FROM 26035KT. TOPS TO FL450.
HAIL TO 1.5 IN...WIND GUSTS TO 60KT POSS.


OUTLOOK VALID 212255-220255
AREA 1...FROM SSM-60NNW YVV-ECK-BUM-SLN-SSM
WST ISSUANCES EXPD. REFER TO THE MOST RECENT ACUS01 KWNS FROM THE
STORM PREDICTION CENTER FOR SYNOPSIS AND METEOROLOGICAL DETAILS.
REF WW 676 677 678.


AREA 2...FROM TTT-MLU-HRV-120SSW LCH-80E BRO-BRO-60SSE
LRD-DLF-TTT
WST ISSUANCES POSS ERLY IN THE PD. REFER TO THE MOST RECENT
ACUS01 KWNS FROM THE STORM PREDICTION CENTER FOR SYNOPSIS AND
METEOROLOGICAL DETAILS.
```

APPENDIX B
CONUS MILITARY BASES AND CORRESPONDING WEATHER FORECAST OFFICES

This is a list of military bases in the conterminous United States (CONUS) and their servicing National Weather Service Weather Forecast Office (WFO). An acronym list for base designations is appended at the end of the military locations list.

Military Location	Weather Forecast Office (WFO)
U.S. Air Force Academy	WFO Colorado Springs, CO
MCLB Albany	WFO Albany, NY
Altus AFB	WFO Norman AFB
Andrews AFB	WFO Sterling, VA
Arnold AFB	WFO Nashville, TN
NAS Atlanta	WFO Atlanta, GA
NSB Bangor	WFO Portland, ME
Barksdale AFB	WFO Shreveport, LA
MCLB Barstow	WFO Fresno, CA
Beale AFB	WFO Sacramento, CA
MCAS Beaufort	WFO Charleston, SC
Bergstrom AFB	WFO San Antonio/Austin, TX
Bolling AFB	WFO Sterling, VA
NS Bremerton	WFO Seattle, WA
NAS Brunswick	WFO Portland, ME
Buckley ANGB	WFO Denver, CO
Camp Fretterd	WFO Sterling, VA
Camp Mabry	WFO Houston, TX
Cannon AFB	WFO Amarillo, TX
Cape Canaveral AFS	WFO Melbourne, FL
Carswell AFB	WFO Fort Worth, TX
Channel Islands ANGS	WFO Los Angeles, CA
Charleston AFB	WFO Charleston, SC
NWS Charleston	WFO Charleston, SC
MCAS Cherry Point	WFO Newport/Moorhead City, NC
Cheyenne Mountain AFS	WFO Colorado Springs, CO
NWS China Lake	WFO Hanford, CA
Columbus AFB	WFO Memphis, TN
NAS Corpus Christi	WFO Corpus Christi, TX
NTTC Corry Station	WFO Mobile, MS
Creech AFB	WFO Las Vegas, NV
NSWCDD Dahlgren	WFO Wakefield, VA

FCTCLANT Dam Neck	WFO Wakefield, VA
Davis-Monthan AFB	WFO Tucson, AZ
Dobbins AFRB	WFO Atlanta, GA
Dover AFB	WFO Philadelphia, PA
Dyess AFB	WFO Fort Worth, TX
Edwards AFB	WFO Harford, CA
Ellsworth AFB	WFO Rapid City, SD
NS Everett, WA	WFO Seattle, WA
Fairchild AFB	WFO Spokane, WA
NAS Fallon	WFO Reno, NV
Forbes Field ANGB	WFO Topeka, KS
Francis E. Warren AFB	WFO Cheyenne, WY
Ft. Belvoir	WFO Sterling, VA
Ft. Benning	WFO Atlanta, GA
Ft. Bliss	WFO El Paso, TX
Ft. Bragg	WFO Raleigh, NC
Ft. Campbell	WFO Nashville, TN
Ft. Carson	WFO Colorado Springs, CO
Ft. Drum	WFO Buffalo, NY
Eglin AFB	WFO Tallahassee, FL
Ft. Eustis	WFO Wakefield, VA
Ft. Huachuca	WFO Tucson, AZ
Ft. Hood	WFO Fort Worth, TX
Ft. Indiantown Gap	WFO State College, PA
Ft. Irwin	WFO Las Vegas, NV
Ft. Knox	WFO Louisville, KY
Ft. Leavenworth	WFO Topeka, KS
Ft. Leonard Wood	WFO Springfield, MO
Ft. Lewis	WFO Seattle, WA
Ft. McPherson	WFO Atlanta, GA
Ft. Polk	WFO Lake Charles, LA
Ft. Riley	WFO Topeka, KS
Ft. Rucker	WFO Tallahassee
Ft. Sill	WFO Lawton, OK
Ft. Stewart	WFO Atlanta, GA
NAS JRB Fort Worth	WFO Fort Worth, TX
Goodfellow AFB	WFO San Angelo, TX
Grand Forks AFB	WFO Fargo, ND
NTC Great Lakes	WFO Chicago, IL
Grissom AFRB	WFO Indianapolis, IN
NCBC Gulfport	WFO Mobile, MS
Hanscom AFB	WFO Boston, MA
Henderson Hall	WFO Sterling, VA
Hill AFB	WFO Salt Lake City, UT
Holloman AFB	WFO El Paso, TX

Homestead AFRB	WFO Miami, FL
Hurlburt AFB	WFO Tallahassee, FL
Hunter AAF	WFO Charleston, SC
NS Ingleside	WFO Corpus Christi, TX
NAS Jacksonville	WFO Jacksonville, FL
MCSA Kansas City	WFO Pleasant Hill, MO
Kelly AFB	WFO San Antonio/Austin, TX
Keesler AFB	WFO Mobile, AL
NAS Key West	WFO Miami, FL
NSB Kings Bay	WFO Jacksonville, FL
NAS Kingsville	WFO Corpus Christi, TX
Kirtland AFB	WFO Albuquerque, NM
Lackland AFB	WFO San Antonio, TX
NAES Lakehurst	WFO Philadelphia, PA
Langley AFB	WFO Wakefield, VA
Laughlin AFB	WFO San Antonio/Austin, TX
Los Angeles AFB	WFO Los Angeles, CA
Camp Lejeune	WFO Wilmington, NC
NAB Little Creek	WFO Wakefield, VA
Little Rock AFB	WFO Little Rock, AR
Los Angeles AFB	WFO Los Angeles,
Luke AFB	WFO Phoenix, AZ
MacDill AFB	WFO Tampa, FL
Malmstrom AFB	WFO Great Falls, MT
March ARB, CA	WFO Los Angeles, CA
Maxwell AFB	WFO Birmingham, AL
NS Mayport	WFO Jacksonville, FL
McChord AFB	WFO Seattle, WA
McConnell AFB	WFO Wichita, KS
McGuire AFB	WFO Philadelphia, PA
NSA Mid-South	WFO Memphis, TN
Minot AFB	WFO Bismarck, ND
MCAS Miramar	WFO San Diego, CA
NAS Meridian	WFO Jackson, MS
Moody AFB	WFO Charleston, SC
Mountain Home AFB	WFO Boise, ID
Naval Post Graduate School	WFO San Franciso, CA
US Naval Academy	WFO Sterling, VA
NSB New London	WFO New York/Upton, NY
Nellis AFB	WFO Las Vegas, NV
NS Newport	WFO Boston, MA
MCAS New River	WFO Newport/Moorhead City, NC
NSA New Orleans	WFO New Orleans, LA
Niagra Falls ARS	WFO Buffalo, NY
NS Norfolk	WFO Wakefield, VA

NSGA Northwest	WFO Wakefield, VA
NAS Oceana	WFO Wakefield, VA
Offutt AFB	WFO Omaha, NE
Onizuka AFS	WFO San Francisco, CA
Otis ANGB	WFO Boston, MA
NCSS Panama City	WFO Tallahassee, FL
MCRD Parris Island	WFO Charleston, SC
NS Pascagoula	WFO Mobile, MS
Patrick AFB	WFO Melbourne, FL
NAS Patuxent River	WFO Sterling, VA
Pease ANGB	WFO Portland, ME
Camp Pendleton	WFO Wakefield, VA
NAS Pensacola	WFO Mobile, FL
Pentagon	WFO Sterling, VA
Peterson AFB	WFO Colorado Springs, CO
NAS Point Mugu	WFO Los Angeles, CA
Pope AFB	WFO Raleigh, NC
NCBC Port Heuneme	WFO Los Angeles, Ca
NS Portsmouth	WFO Gray/Portland, ME
MCB Quantico	WFO Sterling, VA
Randolph AFB	WFO San Antonio/Austin, TX
Robbins AFB	WFO Atlanta, GA
NS San Diego	WFO San Diego, CA
Schriever AFB	WFO Colorado Springs, CO
Scott AFB	WFO St. Louis, MO
Selfridge ANGB	WFO Grand Rapids, MI
Seymour Johnson AFB	WFO Raleigh, NC
Shaw AFB	WFO Charleston, SC
Sheppard AFB	WFO Norman, OK
Simmons AAF	WFO Raleigh, NC
Stennis Space Center	WFO Mobile, MS
Tinker AFB	WFO Norman, OK
Travis AFB	WFO Sacramento, CA
MCAGCC 29 Palms	WFO Las Vegas, NV
Tyndall AFB	WFO Tallahassee, FL
Whiteman AFB	WFO Pleasant Hill, MO
Willow Grove ARS	WFO Philadelphia, PA
Vance AFB	WFO Tulsa, OK
Vandenberg AFB	WFO Los Angeles, CA
Volk Field ANGB	WFO Lacrosse, WI
SCSC Wallops Island	WFO Wakefield, VA
Westfield ANGB	WFO
NAS Whidbey Island	WFO Seattle, WA
Whiteman AFB	WFO Pleasant Hill, MO
NAS Whiting Field	WFO Mobile, FL

NAS JRB Willow Grove	WFO Philadelphia/Mt Holly, PA
Wright-Patterson AFB	WFO Wilmington, OH
NWS Yorktown, VA	WFO Wakefield, VA
MCAS Yuma	WFO Phoenix, AZ

Unit Designations:

AAF	Army Air Field
AFB	Air Force Base
AFRB	Air Force Reserve Base
AFS	Air Force Station
ANGS	Air National Guard Station
FCTCLANT	Fleet Combat Training Center Atlantic
Ft.	Fort
MCAGCC	Marine Corps Air Ground Combat Center
MCAS	Marine Corps Air Station
MCB	Marine Corps Base
MCLB	Marine Corps Logistics Base
MCRD	Marine Corps Recruit Depot
MCSA	Marine Corps Support Activity
NAB	Naval Amphibious Base
NAES	Naval Air Engineering Station
NAS	Naval Air Station
NAS JRB	Naval Air Station Joint Reserve Base
NCBC	Naval Construction Battalion Center
NCSS	Naval Coastal Systems Station
NS	Naval Station
NSA	Naval Support Activity
NSB	Naval Submarine Base
NSGA	Naval Security Group Activity
NSWC	Naval Systems Warfare Center
NTTC	Naval Technical Training Center
NWS	Naval Weapons Station
SCSC	Surface Combat Systems Center

APPENDIX C
CONTINGENCY BACKUP OPERATIONS

Memorandum of Agreement

Between

The National Centers for Environmental Prediction
The National Weather Service
The National Oceanic and Atmospheric Administration

And

The Air Force Weather Agency
The United States Air Force

Regarding

The Backup of Operational Services of the Aviation Weather Center and the Storm Prediction Center by the Air Force Weather Agency at Offutt AFB, NE **(Signed May 2004)**

1. Introduction

1.1 This Memorandum of Agreement (MOA) between the National Centers for Environmental Prediction (NCEP) of the National Weather Service (NWS) of the National Oceanic and Atmospheric Administration (NOAA) and the Air Force Weather Agency (AFWA) of the United States Air Force (USAF) provides for the AFWA to backup the operational services of the Aviation Weather Center (AWC) and the Storm Prediction Center (SPC) and provide an interim site for AWC and SPC forecasters in the event of a prolonged outage at the AWC or SPC respectively. AFWA in turn receives an ancillary data source, the National-Automated Weather Information Processing System (N-AWIPS) from NCEP on a dedicated communications circuit. This capability could be used as a backup source by AFWA for limited NCEP model data in the event of a catastrophic failure of the NWS Telecommunications Gateway or an Information Assurance-driven lockdown of the unclassified data networks at AFWA (DATMS-U, NIPRNET, etc.).

1.2 Specifically, this MOA sets forth a framework for mutually beneficial cooperation, and delineates the respective roles of the AWC, the SPC, and the AFWA in administering the operational backup program. For the purposes of this agreement, the backup of services is defined in National Weather Service Policy Directive (NWSPD) 10-22, December 3, 2003, and in Operations and Services Readiness, National Weather Service Instruction (NWSI) 10-2201, February 5, 2004, entitled Backup Operations. Further, the backup responsibility falls under the Office of the Federal Coordinator for Meteorological Services and Supporting Research's Federal Plan for Cooperative Support and Backup Among Operational Processing Centers FCM-P14-2002.

2. Authority

NWS is authorized to enter into this agreement pursuant to 15 USC 313,49 USC 44720 and 15 USC 1525, the Department of Commerce's joint project authority. AFWA is authorized to enter into this agreement pursuant to DoDI 4000.19 Interservice and Intragovernmental Support. The parties certify that they have a mutual interest in ensuring delivery of the NWS products described herein to the public. The costs associated with this project have been equitably apportioned with each party bearing the costs associated with its participation. This agreement does not involve the transfer of funds between the parties.

3. Agency Responsibilities

3.1 AWC Responsibilities

3.1.1 In the event of any ground system malfunction, security or safety concerns, failure, or degradation to normal AWC operations, the AWC will notify, via Cell phone and Conference Bridge, the AFWA Operations Center (Tel 402-294-2586 option 1) and brief the problem. If deemed necessary by the AWC Lead Forecaster, the AFWA will assume backup operations for the AWC until notified by the AWC that they are prepared to resume normal operations or AWC forecasters arrive at the AFWA. The AWC Lead Forecaster will also notify the Senior Duty Meteorologist (SDM) at NCEP Central Operations (NCO) of all changes in status. AWC will also issue an Area Forecast amendment message notifying users of the outage. The AWC Lead Forecaster will also call the Honolulu Forecast Office and the Tropical Prediction Center to notify them that backup is required for the Flight Information Regions (FIR) in the Pacific and Atlantic Oceans. The AWC Lead Forecaster will notify all parties when the emergency has ended.

3.1.2 If the emergency continues for more than 8 hours, the AWC will send forecasters to AFWA and they will work in a designated area on workstations previously provided to AFWA by the AWC. While stationed at AFWA, the AWC forecasters will resume the full operational tactical program of Significant Meteorological Forecasts (SIGMETs), Convective SIGMETs, Airman's Meteorological Information (AIRMETs), and Area Forecasts.

3.2 SPC Responsibilities

3.2.1 In the event of any ground system malfunction, security or safety concerns, failure, or degradation to normal SPC operations, the SPC will notify, via telephone, the AFWA Operations Center (Tel 402-294-2586 option 1) and brief the problem. If it is deemed necessary by the SPC Lead Forecaster, the AFWA will assume backup operations for the SPC until notified by the SPC that they are prepared to resume operations or SPC forecasters arrive at the AFWA. The SPC Lead Forecaster will also notify the SDM at NCO of all changes in status.

3.2.2 If the backup is expected to continue for more than 24 hours, the SPC will send forecasters to the AFWA. They will work in a designated area on workstations provided by the AFWA. The SPC Lead Forecaster will also notify the SDM at NCO of all changes in status.

3.3 AFWA Responsibilities

3.3.1 Once notified by the AWC to begin backup operational support, the AFWA will provide, as required, the following products to the appropriate NWS customers until AWC forecasters arrive at the AFWA or the backup is terminated.

3.3.1.1 Convective SIGMETs, Non-Convective SIGMETs, and AIRMETs as defined in NWSI 10-811.

3.3.1.2 AFWA will provide additional space for AWC forecasters to carryout Area Forecast (FA) backup responsibility.

3.3.2 Once notified by the SPC to start operational backup, the AFWA Lead Meteorologist would call the SPC to acknowledge and confirm assumption of operational backup. Upon assumption of operational backup, the AFWA will provide, as required, the following products to the appropriate NWS customers until SPC forecasters arrive at the AFWA or the backup is terminated.

3.3.2.1 Day 1/Day 2/Day 3 Convective Categorical Outlooks (text and graphics).

3.3.2.2 Weather Watches (text and graphics).

3.3.3 The AFWA will provide for necessary maintenance on appropriately designated computer hardware located at the AFWA to properly run all required software for AWC/SPC backup operations. The AFWA Systems and Network Management Branch (SCHS) will also provide access to the NWS equipment for AWC/SPC technical personnel to install and maintain operating software.

3.3.4 The AFWA will ensure there is a fully trained cadre of forecasters to provide AWC/SPC backup operations. This will include travel by AFWA Strategic Center Backup Lead Meteorologist to the AWC/SPC biannually for training. The AFWA shall be solely responsible for all costs associated with such travel to AWC/SPC.

3.3.5 The AFWA will be responsible for providing technology refresh, upgrades, and sustainment to all AFWA owned hardware.

3.4 NWS/NCEP Responsibilities

3.4.1 The NCEP/NCO SDM, upon notification by either SPC or AWC that AFWA has assumed backup operations, will issue a notification message of a change in status that will be transmitted on Automated Weather Information Processing System (AWIPS) and sent to TOC/Gateway for distribution to all other appropriate communication circuits.

3.4.2 The NCEP will provide appropriate data sets (e.g., model output, satellite, observations, pireps, etc.) for use on appropriately designated workstations at the AFWA. In addition, training and training materials will be provided as required on system hardware and software.

3.4.3 The NCEP will provide software support, installation, and maintenance to the AFWA for all backup software requirements, including software upgrades for AWC/SPC unique applications and product generation software.

3.4.4 The NCEP will be responsible for providing technology refresh, upgrades, and sustainment to all NWS hardware located at the AFWA.

3.4.5 The NCEP will be solely responsible for all costs associated with travel to AFWA for AWC/SPC technology exchange visits.

3.5 Joint NWS/AFWA Responsibilities

3.5.1 AFWA/XOG and NCEP (AWC and/or SPC) will maintain a mutually coordinated and agreed upon listing of equipment (computer hardware and software), floor plans, communication circuits, etc. associated with this operational backup support. Listing will also document ownership of the equipment.

3.5.2 AFWA and NCEP will share equally the costs incurred for the communications link between NCEP and AFWA.

3.5.3 NWS and AFWA will conduct scheduled practice backup exercises on a quarterly basis. (Request AWC/SPC personnel present)

4. Additional Terms and Conditions

4.1 NWS and AFWA participation in this agreement is subject to the availability of appropriated resources. Nothing herein is intended to conflict with current NOAA and USAF directives. If any terms of this agreement are inconsistent with directives of either of the agencies entering into the agreement, then those portions of the agreement that are inconsistent shall be held invalid. Remaining terms and conditions not affected by the inconsistency shall continue in full force and effect.

4.2 Should a disagreement arise in interpretation of the provisions contained herein, or of any amendment or revision, the parties shall attempt to reconcile the differences first at the operating level. Each party for consideration shall state the areas of disagreement in writing. If the agreement cannot be reached in thirty days, the disputing parties shall forward the written presentation and documents relating to the disagreements to respective higher officials for appropriate resolution.

5. Effective Date, Period of Performance, and Termination

5.1 The procedures outlined in this MOA are detailed in NWS directives (Operations Manual Chapters and associated Operations Manual Letters). These procedures will be implemented beginning on the effective date of the update(s) to those directives that reflect this MOA. This MOA becomes effective on the date of the final signature and shall remain in effect until terminated.

5.2 This MOA shall remain in effect for 5 years from the latest date appearing below and shall be reviewed at least annually by all parties concerned to ensure completeness and accuracy. At such time, and at any other time agreed to by the parties, this agreement may be modified as appropriate, and mutually agreed. This MOA may be terminated by either party, but shall require 180 days advance written notification by the terminating party.

APPENDIX D
STATE CONTACTS FOR NAWAS NETWORK

Alabama
Fred Springall
205-834-1375
Alabama Emergency Management Agency
5898 County Road 41
Clanton, AL 35045-5160

Arizona
Harry E. Border, Program Director
602-231-6214
Arizona Division of Emergency Services
5626 East McDowell Road
Phoenix, AZ 85008

Arkansas
Jim Stalnaker
501-329-5601
Arkansas Office of Emergency Services
P.O. Box 758
Conway, AK 72032

California
Lloyd Darrington
916-427-4375; Fax: 916-427-1677
California Warning Center
State Office of Emergency Services
2800 Meadowview Road
Sacramento, CA 95832

Colorado
Dave Holms
303-273-1619
Colorado Disaster Emergency Services
1500 Golden Road
Camp George West
Basement Room D-18
Golden, CO 80401

Connecticut
Tom Walsh
860-566-4737
Office of Emergency Management
360 Broad Street
Hartford, Connecticut 06105

District of Columbia
Bill Curry
202-673-7353
2000 14th St., NW 8th Floor
Washington, D.C.

Delaware
Alan McClements
302-834-4531
State Of Delaware
Department of Public Safety
Emergency Planning and Operations Division
P.O. Box 527
Delaware City, Delaware 19706

Florida
John Fleming
904-448-1900; FAX: 904-448-6250 (to get updated records: 904-448-1320)
Division of Emergency Management
2740 Centerview Drive
Rhyne Building - Room #175
Tallahassee, FL 32399

Georgia
Mike Taylor
404-624-7222
Georgia Emergency Management Agency
P.O. Box 18055
Atlanta, GA 30316-0055

Idaho
Vicki Miller, Communications & Resource Officer
208-334-3460
Idaho Bureau of Disaster Services
650 West State Street
Boise, Idaho 83720

Illinois
John Myers
217-782-4602
Illinois Emergency Management Agency
110 East Adams Street
Springfield, IL 62706

Indiana
Mike Rollins
317-233-6055
Indiana State Police
Indiana State Office Building
100 North Senate Avenue
Indianapolis, IN 46204

Iowa
Dick Clement
515-281-3231
Iowa Office of Disaster Service
Hoover State Office Building, Level A
Des Moines, Iowa 50319

Kansas
Larry Waters, Communications Operator II
913-296-13102/6908
State of Kansas; The Adjutant General
Division of Emergency Preparedness
P.O. Box C-300
Topeka, Kansas 66601-0300

Kentucky
Omar Marshall
502-564-8617
Kentucky Emergency Management Agency
Boone Center
Frankfort, KY 40601-6168

Louisiana
John Buie & Jim Wilkes
504-342-5470
State of Louisiana Military Department
Office of Civil Defense & Emergency Preparedness
P.O. Box 44217
Baton Rouge, LA 70804

Maine
Joe Grimmig
207-626-4503
State Civil Defense Control Center
State House
72 State Office Building
Augusta, Maine 04333

Maryland
Hank Black, Communications Officer
877-636-2872
Maryland Emergency Management Agency
2 Sudbrook Lane East
Pikesville, MD 21208

Massachusetts
Steve Finks
508-820-2020
Massachusetts Emergency Management Agency
400 Worchester Road
Framingham, MA 01701

Michigan
Tom Newell
517-334-5126/5026
Michigan State Police
Emergency Management Division
Knapps Center, Suite 300
300 South Washington Square
Lansing, Michigan 48913

Minnesota
Tom Cherney
612-296-0455
Division of Emergency Management
B-5 State Capital
St. Paul, Minnesota 55155

Mississippi
Alvin Reynold
601-960-9000
Mississippi Emergency Management Agency
P.O. Box 4501
Jackson, Mississippi 39296-4501

Missouri
Cathy Zumwalt
573-526-9146
Missouri Emergency Management Agency Communications Center
1717 Industrial Drive
Jefferson City, MO 65101

Montana
Homer Young, Communications Officer
406-444-6911
Emergency Management Agency
P.O. Box 4789
Helena, Montana 59604

Nebraska
Bob Eastwood
402-471-7414
Communications Manager
Nebraska State Civil Defense Agency
1300 Military Rd.
Lincoln, NE 68508-1070

Nevada
Bob Minter
702-687-4240
Nevada Division of Emergency Management
2525 South Carson Street
Carson City, NV 89710

New Hampshire
Chuck Welch
603-271-2231
Office of Emergency Management
State Office Park South
107 Pleasant Street
Concord, NH 03301

New Jersey
Bob Schroeder
609-530-6019
Division of State Police
Emergency Management Section
P.O. Box 7068
West Trenton, NJ 08628-0068

New Mexico
Larry Austin
505-827-9242
New Mexico State Police Headquarters
4491 Cerillus Road
P.O. Box 1628
Santa Fe, NM 87504

New York
Earl Dressel
518-457-2200
Communications Section
NY State Emergency Management Office
Building 22 - State Campus
Albany, NY 12226-5000

North Dakota
Larry Ruebel
701-328-8108
Division of Emergency Management
P.O. Box 5511
Bismarck, ND 58502-5511

North Carolina
Clay Benton
919-715-4264
North Carolina State EOC
116 West Jones Street
Raleigh, NC 27603

Ohio
Dean Bolton
614-889-7155
Emergency Management Agency
2825 West Granville Road
Columbus, Ohio 43235-2712

Oklahoma
Ron Hill
405-521-2481
Oklahoma Civil Emergency Management Agency
P.O. Box 53365
Oklahoma City, OK 73152

Oregon
Joseph Cunningham
503-378-6377
Oregon Emergency Management
595 Cottage Street N.E.
Salem, OR 97310

Pennsylvania
Steve Vergot
717-651-2001/2039
Pennsylvania State
Council of Civil Defense
Foster and Commonwealth
Harrisburg, PA 17105-3321

Rhode Island
Jim Bell
401-946-9996
Emergency Management Agency
State House
Room 27
Providence, RI 02903-1197

South Dakota
Jim Ward
605-773-6425
Department of Military & Veterans Affairs
500 East Capitol
Pierre, SD 57501

South Carolina
Kenton Towner
803-734-8020
South Carolina EOC
1429 Senate Street
Columbia, SC 29200

Tennessee
Lynn Richland & James Barrett
615-741-0008
Tennessee Emergency Management Agency
Emergency Operations Center
3041 Sidco Drive
P.O. Box 41502
Nashville, TN 37204-1502

Texas
Linda Moore
512-424-2278
Texas Department of Public Safety
5805 North Lamar Blvd.
P.O. Box 4087
Austin, TX 78773-0001

Utah
Jim Brown
801-538-3400
State of Utah
Department of Public Safety
Division of Comprehensive Emergency Management
State Office Building, Room 1110
Salt Lake City, UT 84114

Vermont
Chris Fuhrmeister
802-244-8727
Emergency Management Department
103 South Main Street
Waterbury, VT 05671

Virginia
Ken Crumpler
804-897-6606
Commonwealth of Virginia
Department of Emergency Services
310 Turner Road
Richmond, VA 23225

Washington
Jimmie Hocutt, Communications Officer
206-459-9191
State of Washington
4220 East Martin Way
Olympia, WA 98504

West Virginia
Clay Carney, Communications Officer
304-348-5380
State of West Virginia
Office of Emergency Service
Main Capitol, Room EB-80
Charleston, WV 25305

Wisconsin
Alan Wohlfred, Director, Communications & Warning
608-266-3232
Wisconsin Division of Emergency Government
4802 Sheboygan Ave, Room 99A
P.O. Box 7865
Madison, WI 53707-3232

Wyoming
Bill Morton
307-777-4905
Wyoming Emergency Management Agency
Basement Emergency Operations Center
5500 Bishop Blvd
P.O. Box 1709
Cheyenne, WY 8200

APPENDIX E
ABBREVIATIONS AND ACRONYMS

-A-

ADWS	Automatic Digital Weather Switch
AFB	Air Force Base
AFRB	Air Force Reserve Base
AFRL	Air Force Research Laboratory
AFS	Air Force Station
AFSS	Automated Flight Service Station
AFW	Air Force Weather
AFWA	Air Force Weather Agency
AIRMET	Airmen's Meteorological Information
ALDARS	Automated Lightning Detection and Reporting System
AM	Amplitude Modulation
AMIS	Advance Meteorological Information System
AMOS	Automated Meteorological Observing Stations
AMSU	Advanced Microwave Sounding Unit
ANGB	Air National Guard Base
APT	Automatic Picture Transmission
ARTCC	Air Route Traffic Control Center
ASOS	Automated Surface Observing System
ASR	Airport Surveillance Radar
ATCT	Air Traffic Control Tower
ATM	Asynchronous Transfer Mode
ATOVS	Advanced TIROS-N Operational Vertical Sounder
AUTODIN	Automatic Digital Network
AWC	Aviation Weather Center
AWIPS	Advanced Weather Interactive Processing System
AWOS	Automated Weather Observing System
AVHRR	Advanced Very High Resolution Radiometer

-B-

BLM	Bureau of Land Management

-C-

C^{4}I	Command, Control, Communications, Computers, and Intelligence
CDDF	Central Data Distribution Facility
C/ESORN	Committee for Environmental Services, Operations and Research Needs
CIRA	Cooperative Institute for Research in the Atmosphere
C-MAN	Coastal Marine Automated Network
COMET	Cooperative Program for Operational Meteorology, Education and Training
CONUS	Continental United States
CRS	Console Replacement System (NWS)
CWS	Combat Weather Squadron
CWSU	Center Weather Service Unit
CWA	Center Weather Advisory

-D-

DDS	Domestic Data Service
DCO	Data Collection Offices
DET	Detachment
DMSP	Defense Meteorological Satellite Program
DOC	Department of Commerce
DOD	Department of Defense
DOT	Department of Transportation
DSN	Defense Switching Network

-E-

EAS	Emergency Alert System
EMWIN	Emergency Managers Weather Information Network
EOC	Emergency Operations Center

-F-

FAA	Federal Aviation Administration
FCC	Federal Communication Commission
FCMSSR	Federal Committee for Meteorological Services and Supporting Research
FCM	Federal Coordinator for Meteorology
FEMA	Federal Emergency Management Agency
FHWA	Federal Highway Administration
FM	Frequency Modulation
FNMOC	Fleet Numerical Meteorology and Oceanography Center
FO	Military Weather Advisory Future Outlooks
FOS	Family of Services
FSS	Flight Service Station
FTS	Federal Telecommunications Service

-G-

GAC	Global Area Coverage 4 km Resolution
GMDSS	Global Maritime Distress and Safety System
GOES	Geostationary Operational Environmental Satellite
GPS	Global Positioning System
GTS	Global Telecommunications System
GVAR	GOES Variable

-H-

H+55	55 minutes past the Hour
HAWCNET	High-speed Asynchronous transfer mode Weather Communications NETwork
HAZUS	HAZards U.S.
HAZUS_MH	HAZards U.S. Multi-Hazard
HNL	Honolulu, HI WFO
HQ	Headquarters
HRPT	High Resolution Picture Transmission
HF	High Frequency

-I-

IAP	International Airport
ICMSSR	Interdepartmental Committee for Meteorological Services and Supporting Research
IMO	International Maritime Organization
IR	Infrared
ITWS	Integrated Terminal Weather System
IWIN	Interactive Weather Information Network

-J-

JAAWIN	Joint Air Force and Army Weather Information Network
JAG/SLSO	Joint Action Group for Severe Local Storms Operations
JET	Joint Environmental Toolkit
JOTS	Joint Operational Tactical System
JPDO	Joint Planning and Development Office

-L-

LAC	Local Area Coverage 1.1 km resolution
LDS	Lightning Detection System
LF	Light Fine Video Data (1/3 nm (0.6 km))
LOA	Letter of Agreement
LS	Light Smooth Video Data (1.5 - 2.0 nm (2.8 - 3.7 km))

LVL	Level

-M-

MDR	Manually Digitized Radar
METAR	aviation routing weather report
METOC	Meteorology and Oceanography (as in commands, centers, or DETs) or Meteorological and Oceanographic (as in data or products)
METSAT	Meteorological Satellite
METWATCH	Meteorological Watch
MF	Medium Frequency
MIC	Maximum Instantaneous Coverage
MIDDS	Meteorological Integrated Data Display System
MSU	Microwave Sounding Unit
MWA	Military Weather Advisory

-N-

NASA	National Aeronautics and Space Administration
NAVMETOCCOM	Naval Meteorology and Oceanography Command
NAVTEX	a primary GMDSS
NAWAS	National Warning System
NCEP	National Centers for Environmental Prediction
NCF	Network Control Facility
NCO	NCEP Central Operations
NCWF	National Convective Weather Forecast
NDBC	National Data Buoy Center
NESDIS	National Environmental Satellite, Data, and Information Service
NEXRAD	Next Generation Radar (WSR-88D)
NGDC	National Geophysical Data Center
NHC	National Hurricane Center
NIMA	National Imagery Mapping Agency
NIPRNET	Non-secure Internet PRotocol NETwork
NMOSS	Navy Mobile METOC Support System
NOAA	National Oceanic and Atmospheric Administration
NOAAPort	NOAA communications system for data and products
NODDS	Navy Oceanographic Data Distribution System
NPHs	Natural Phenomena Hazards
NRC	Nuclear Regulatory Commission
NRP	National Response Plan
NSLSOP	National Severe Local Storms Operations Plan
NSSL	National Severe Storms Laboratory
NWP	Numerical Weather Prediction
NWR	NOAA Weather Radio
NWS	National Weather Service
NWSTG	NWS Telecommunications Gateway

NWWS   NOAA Weather Wire Service

-O-

OBS	observations
OFCM	Office of the Federal Coordinator for Meteorological Services and Supporting Research
OJCS	Organization of the Joint Chiefs of Staff
OLS	Operational Line Scanning
OWS	Operational Weather Squadron

-P-

PATWAS	Pilots Automatic Telephone Weather Advisory Service
PIBAL	Pilot Balloon
PIREP	Pilot Report
POES	Polar Orbiter Environmental Satellite
PPS	Public Product Service
PWW	Point Weather Warnings

-R-

RAWS	Remote Automatic Weather Station
R&D	Research and Development
RAMSDIS	Regional and Mesoscale Meteorology Team Advanced Meteorological Satellite Demonstration and Interpretation System
RAWIN	Rawinsonde
RCM	Radar Coded Message
RFC	River Forecast Center
RMTN	Regional Meteorological Telecommunications Network
RPG	Radar Product Generator
ROB	Automated Radar Observation (WSR-88D)
RTOVS	Revised TIROS-N Operational Vertical Sounder
RSO	Rapid Scan Operations
RVR	Runway Visual Range

-S-

SAB	Satellite Analysis Branch
SafetyNET	a primary GMDSS
SAME	Specific Area Message Encoders
SAWRS	Supplementary Aviation Weather Reporting Station
SD	Storm Detection
SDHS	Satellite Data Handling System
SDM	Senior Duty Meteorologist at NCEP
SIGRAD	Significant Radar Message
SIGMET	Significant Meteorological Information

SLSO	Severe Local Storms Operations
SNOINCR	Snow Increasing Rapidly report
SNOTEL	Snow Telemetry
SOLAS	Safety of Life at Sea
SPC	Storm Prediction Center
SPECI	aviation selected special weather report
SPP	Shared Processing Program
SR	Stored Data
SRC	State Relay Center
SRSO	Super Rapid Scan Operations
SSCs	Structures, Systems, and Components
SSM/I	Special Sensor Microwave Imager
SSM/T-1	Special Sensor Microwave Temperature Sounder
SSM/T-2	Special Sensor Microwave Moisture Sounder
SSU	Stratospheric Sounding Unit

-T-

TAA	Total Area Affected
TDWR	Terminal Doppler Weather Radar
TF	Thermal Fine Data (1/3 nm (0.6 km))
TIROS	Television Infrared Observation Satellite
TOVS	TIROS Operational Vertical Sounder
TPC	Tropical Prediction Center
TRACONS	Terminal Radar Approach Controls
TS	Thermal Smooth Data (1.5 - 2.0 nm (2.8 - 3.7 km))
TWEB	Transcribed Weather Broadcast

-U-

UA	Routine pilot reports
USA	United States Army
USAF	United States Air Force
USCG	United States Coast Guard
USMC	United States Marine Corps
USN	United States Navy
UTC	Universal Coordinated Time
UUA	Urgent pilot reports

-V-

VAS	VISSR Atmospheric Sounder
VHF	Very High Frequency
VHRR	Very High Resolution Radiometer
VIP	Video Integrated Processor
VISSR	Visible Infrared Spin Scan Radiometer

VOR   VHF Omni-Directional Radio Range
VOS   Voluntary Observing Ship

-W-

WFO   Weather Forecast Office
WIBIS   Severe Weather Watch Will Be Issued
WMO   World Meteorological Organization
WMSCR   Weather Message Switching Center Replacement
WSOM   Weather Service Operations Manual

APPENDIX F
MEMBERS OF THE JOINT ACTION GROUP FOR SEVERE LOCAL STORMS OPERATIONS

Mr. John Ferree, Chair
NOAA/National Weather Service

Mr. Chris Maier
NOAA/National Weather Service

Mr. Ron Olson
NOAA/National Weather Service

Mr. Jay Hanna
NOAA/NESDIS

Mr. Brian Hughes
NOAA/NESDIS

Mr. Jonathan Berkson
DHS/United States Coast Guard

Mr. Kurt Nelson
Department of Defense/Navy

Mr. Rickey Petty
Department of Energy

Mr. Victor Passetti
DOT/Federal Aviation Administration

Paul Pisano
DOT/Federal Highway Administration

Mr. Don Eick
National Transportation Safety Board

Mr. Greg Carbin
NOAA/National Weather Service

Mr. George Serafino
NOAA/ National Environmental Satellite, Data, and Information Service

Mr. Dan Catlett
Federal Emergency Management Agency

Mr. Paul Seymour
NOAA/NESDIS

SMSgt Craig Kirwin
Department of Defense/Air Force

Mr. Robert Mason
DOI/USGS

Mr. Ray Murphy
DOT/Federal Highway Administration

Ms. Charlene M. Wilder
DOT/Federal Transit Administration

Ms. Jocelyn Mitchell
Nuclear Regulatory Commission

DONELL WOODS, Executive Secretary
Office of the Federal Coordinator for Meteorological Services and Supporting Research